Do Brilliantly

A2 Maths

John Berry, Ted Graham
& Roger Fentem
Series Editor: Jayne de Courcy

Published by HarperCollins*Publishers* Limited
77–85 Fulham Palace Road
London W6 8JB

www.**Collins**Education.com
On-line support for schools and colleges

First published 2002

ISBN 0 00 712433 3

British Library Cataloguing in Publication Data
A catalogue record for this book is available from the British Library

Edited by Joan Miller
Production by Kathryn Botterill
Design by Gecko Ltd
Cover design by Susi Martin-Taylor
Printed and bound by Scotprint

Acknowledgements
The Author and Publishers are grateful to the following for permission to reproduce copyright material in the following questions:

- p.12 Q1, p.12 Q2, p.14 Q2, p.18 Q1, p.18 Q2, p.25 Q2, p.34 Q2, p.38 Q2 (worked example), p.38 Q1, p.39 Q6, p.43 Q5, p.47 Q1, p.48 Q4, p.51 Q2, p.52 Q7, p.56 Q3, p.60 Q2, p.60 Q5, p.64 Q2 reproduced with the permission of the Assessment and Qualifications Alliance **AQA**
- p.14 Q6 reproduced with the permission of the Assessment and Qualifications Alliance **AQA (AEB)**
- p.34 Q6 reproduced with the permission of the Assessment and Qualifications Alliance **AQA (SEB)**
- p.10 Q7, p.14 Q7, p.18 Q6, p.22 Q1, p.22 Q7, p.26 Q3, p.32 Q1, p.38 Q2, p.43 Q3, p.47 Q2, p.48 Q3, p.51 Q1, p.52 Q5, p.56 Q2, p.60 Q1, p.65 Q4 reproduced with the permission of **EDEXCEL**
- p.14 Q4 reproduced with the permission of **EDEXCEL (ULEAC)**
- p.10 Q3, p.10 Q6, p.14 Q1, p.14 Q5, p.18 Q7, p.22 Q2, p.22 Q3, p.24 Q1, p.26 Q2, p.34 Q1, p.36 Q1, p.39 Q3, p.43 Q2, p.48 Q5, p.48 Q6, p.48 Q7, p.52 Q6, p.56 Q1, p.60 Q4, p.64 Q1, p.65 Q5 reproduced with the permission of Oxford Cambridge RSA Examinations **OCR/MEI/UCLES**
- p.34 Q4 reproduced with the permission of the Welsh Joint Education Committee/Cyd-bwyllgor Addysg Cymru **WJEC**.

Note:
The Awarding Bodies listed above accept no responsibility whatsoever for the accuracy or method of working in the answers given, which are solely the responsibility of the authors and publishers.

Illustrations
Cartoon artwork – Roger Penwill
DTP artwork – Gecko Ltd

Contents

How this book will help you

by John Berry, Ted Graham and Roger Fentem

Exam practice – how to answer questions better

This book will help you to improve your performance in your A2 Mathematics exam.

Your A2 Maths exams, like all A2 exams, will include **synoptic assessment**. This means that questions will be set which test your knowledge and understanding of all aspects of the specification. To perform well in these questions, **you must be familiar with all the mathematics that you have studied for your AS exams, as well as what you have learned for your A2 examinations**. However, in mathematics this is not a big problem because, most of the time, the new mathematics that you have learned for A2 builds on the mathematics that you have already studied. For example, a question on trigonometric identities may require you to solve a simple trigonometric equation which you will have covered at AS.

Each chapter in this book is broken down into four separate elements, aimed at giving you as much practice and guidance at answering questions as possible.

1 'Key points to remember' and 'Don't make these mistakes'

On these pages we outline the key points that you need to know for each of the topics covered. These should not be new to you – the aim is to provide **a reminder that will help to refresh your memory**. We also include **lists of formulae that you need to know**. In the A2 exam you are expected to learn quite a lot of formulae rather than being able to look them up in a formula book. It is very important to know which formulae you need to learn and which will be provided in the examination. We also include **a list of common mistakes that you must try to avoid**.

2 Exam Questions and Answers with 'How to score full marks'

We have used a number of exam-type questions to **illustrate the methods that you will be required to use in your examination**. In each case we provide the question, followed by a sample correct solution. Alongside this **we have provided notes to guide you through the solution. We have also indicated the steps needed to score full marks**.

3 Questions to try

This is where **you get the opportunity to practise answering the types of question that you will meet in your exam**. This section starts with some easier questions to warm you up and then includes actual exam questions or exam-style questions. **You need to attempt all the appropriate questions as this sort of practice will help you to reach your full potential in your exam**.

4 Answers to the Questions to try

At the back of the book you will find **solutions to all of the Questions to try**. Try answering each question first, then look at the answer if you are really stuck or when you think that you have completed the question. **Alongside each solution we have written a commentary that identifies the key stages in the solution – this will help you if you get stuck**. The commentary will also help you to make sure that you have not omitted any important working that you should have shown.

This book focuses on the most important or difficult parts of the A2 mathematics core, which all exam boards' specifications must cover for an A2 award. The table below shows how the topics in this book fit into the modules for the different examination boards. (Some sections may contain a little more material than is required for your examination.) If you need extra practice in the AS topics, you will find material in *Collins Do Brilliantly AS Maths*.

Chapters in this book	AQA–A	AQA–B	EDEXCEL	OCR–A	OCR–B (MEI)
1 The binomial expansion	P3	P4/P5	P2/P3	P2/P3	P3
2 Trigonometry	P2	P4/P5	P2	P3	P3
3 Differentiation techniques	P2/P3	P4/P5	P3	P3	P2/P3
4 Integration	P2/P3	P4/P5	P3	P3	P2/P3
5 Vectors	P3	P5	P3	P3	P3
6 Proof	P1/P2/P3	P1/P2/P4/P5	P1/P2/P3	P1/P2/P3	P1/P2/P3
7 Coordinate geometry	P2	P4	P3	P3	P1/P3
8 Differential equations	P3	P5	P3	P3	P3
9 Partial fractions and integration	P3	P4	P3	P3	P3
10 Moments and centres of mass	M2	M1	M2	M2	M2
11 Energy	M2	M2+M3	M2	M2	M2
12 Calculus in mechanics	M1/M2	M2+M3	M2	M1	M1
13 Projectiles	M1	M1	M2	M2	M1
14 Momentum and the coefficient of restitution	M2	M4	M2	M2	M2
15 The Poisson distribution	S2	S1	S2	S2	S2
16 The chi-squared (χ^2) distribution	S2	S2/S4	S3	S3	S3/S4
17 Hypothesis testing	S2	S2	S2	S2	S1/S2
18 Approximating distributions	S1/S2	S1	S2	S2/S3	S2
19 Continuous random variables	S1/S3	S2/S4	S2/S5	S2/S4	S2/S3

Proof

Proof is a part of all the pure mathematics modules that you will be required to take. The language and principles of proof are introduced at AS level, while the techniques for proof are covered at A2. These techniques are used in almost every mathematics solution, as the steps build progressively towards producing the answer (see page 13 Q3b). You should see this demonstrated in the solutions, especially those provided in the chapters and at the end of the book for the pure mathematics modules.

- There are two types of marks available to you: **method marks and accuracy marks**. Accuracy marks are awarded for correct working and answers. Method marks are awarded for working that is of the type required, but that contains a reasonably minor error.

  For example, if, when solving a quadratic equation, you fail to include a negative sign with one of the numbers, you would lose an accuracy mark, but may be awarded a method mark because you have shown that you know how to solve a quadratic equation. **In order to gain method marks it is very important that you show your working clearly** so that the examiner can give you credit if the answer is wrong. Short comments to explain to the examiner what you are trying to do may help.
- **In 'show that…' questions, make sure that you really do show that the result is true**. You should include all the necessary steps in your working. Students can miss out obvious – but important – steps when answering these types of question.
- **Make sure that you have learned the formulae that you need to know for each exam**. You may like to make a list that you can spend time studying. Imagine sitting in the exam looking at a GP, but being unable to remember the formula that you need to find its sum. Also be aware of which formulae are in the formula book for your exam board and be able to find them when you need them.
- **Be aware of which examinations for your exam board have calculator restrictions**. This will apply to some of the pure maths papers. You should know which calculators to take to which exam. A graphics calculator can be an advantage in an examination, but you do need to have spent time getting to know how to use it.
- When you are asked for an exact answer or are asked to

  'show that $x = \frac{1}{\sqrt{3}}e^{\sqrt{3}\pi}$'

  or similar, **do not use decimal approximations as you are working**. Always work with surds or fractions until you obtain the required result. If the required answer includes surds, π or e, you should not convert.
- **Look at the number of marks that are awarded for each part of a question.** This will give you a guide to how much work you have to do. A question that has one mark for an explanation or comment will expect one sentence. Do not write half a page of explanation, as you will be wasting time.
- **Use the time that you have available in the examination productively.** If you know that you are not progressing with a question, stop and move on to another one. Also always try the later parts of questions, even if you cannot do the first part. Exam questions often use the 'show that…' format so that students can attempt later parts even if they can't do the 'show that…' part.

1 The binomial expansion

Key points to remember

- The **binomial expansion** is used to expand expressions of the form $(a \pm b)^n$.
- When n is a **positive integer**, the expansion contains a finite number of terms as given in the formula below.

  $$(a+b)^n = a^n + \binom{n}{1}a^{n-1}b + \binom{n}{2}a^{n-1}b^2 + \ldots + \binom{n}{r}a^{n-r}b^r + \ldots + b^n \qquad n \in N$$

  where $\binom{n}{r} = {}^nC_r = \dfrac{n!}{(n-r)!r!}$

  This can be found in your formula book. Make sure that you know where it is.
- When n is a **real number** but not a positive integer, the expansion will give an infinite number of terms. You will normally be asked to calculate the first few. You can do this, using the result below, which can be found in your formula book.

  $$(1+x)^n = 1 + nx + \frac{n(n-1)}{2}x^2 + \ldots + \frac{n(n-1)(n-2)\ldots(n-r+1)}{r!}x^r + \ldots$$

  $$|x| < 1, x \in \mathfrak{R}$$
- Remember that the expansion is only valid for $|x| < 1$ when n is not a positive integer.

Formulae you must know

All of the key formulae for this topic will be included in your formula book. Make sure that you can find and use them.

Don't make these mistakes ...

- Don't forget to include the – signs when there is one in the bracket.
- Don't make arithmetic errors when simplifying expansions.
- In cases like $(1 + 3x)^n$, don't forget to use $(3x)^2 = 9x^4$ etc. when doing the expansion.
- Don't forget to consider when an expansion will be valid if n is not a positive integer.

Exam Questions and Answers

Q1 Expand $(2 + \frac{x}{2})^4$.

$$\left(2 + \frac{x}{2}\right)^4 = 2^4 + \binom{4}{1} \times 2^3 \times \left(\frac{x}{2}\right) + \binom{4}{2} \times 2^2 \times \left(\frac{x}{2}\right)^2 + \binom{4}{3} \times 2 \times \left(\frac{x}{2}\right)^3 + \left(\frac{x}{2}\right)^4$$

$$= 2^4 + 4 \times 2^3 \times \left(\frac{x}{2}\right) + 6 \times 2^2 \times \left(\frac{x}{2}\right)^2 + 4 \times 2 \times \left(\frac{x}{2}\right)^3 + \left(\frac{x}{2}\right)^4$$

$$= 16 + 4 \times 8 \times \left(\frac{x}{2}\right) + 6 \times 4 \times \left(\frac{x}{2}\right)^2 + 4 \times 2 \times \left(\frac{x}{2}\right)^3 + \left(\frac{x}{2}\right)^4$$

$$= 16 + 16x + 6x^2 + x^3 + \frac{x^4}{16}$$

Q2 If $(1 + ax)^n = 1 + 15x + 90x^2 + kx^3 + \ldots$, find a, n and k.

$$(1 + ax)^n = 1 + n(ax) + \frac{n(n-1)}{2}(ax)^2 + \frac{n(n-1)(n-2)}{6}(ax)^3 + \ldots$$

$$= 1 + nax + \frac{n(n-1)}{2}a^2x^2 + \frac{n(n-1)(n-2)}{6}a^3x^3 + \ldots$$

The coefficient of x is 15.

$an = 15$ or $a = \frac{15}{n}$

The coefficient of x^2 is 90 so

$$\frac{n(n-1)}{2}a^2 = 90$$

How to score full marks

- Use the formula that is quoted above for expanding $(a + b)^n$. Note that in this case $a = 2$ and $b = \frac{x}{2}$. The whole of the term $\frac{x}{2}$ is squared, cubed, etc.
- Take care when simplifying the expression.
- You usually write the expansion in increasing powers of x.
- First expand $(1 + x)^n$ using the formula and replacing x by ax.
- Compare the coefficients of x in both sides, to form an equation relating a and n.
- Compare the coefficients of x^2 to form a second equation.

Substituting for a gives:

$$\frac{n(n-1)}{2} \times \left(\frac{15}{n}\right)^2 = 90$$

$$n - 1 = \frac{90 \times 2}{15^2} n$$

$$n - 1 = \frac{4}{5} n$$

$$\frac{1}{5} n = 1$$

$$n = 5$$

Then $a = \frac{15}{n} = \frac{15}{5} = 3$

and $k = \frac{n(n-1)(n-2)}{6} a^3$

$= \frac{5 \times 4 \times 3}{6} \times 3^3$

$= 270$

How to score full marks

- Substitute for a in the second equation, using $a = \frac{15}{n}$ and solve for n.
- Use $n = 5$ to find a.
- Substitute the values of a and n in the expansion to find k, the coefficient of x^3.

Q3 **(a)** Expand $(1 + 4x)^{\frac{1}{2}}$ in ascending powers of x up to and including x^3.

(b) State the range of values of x for which the expansion is valid.

(c) Use the expansion to find $\sqrt{1.4}$, correct to 3 significant figures.

(a) $(1 + 4x)^{\frac{1}{2}} = 1 + \frac{1}{2} \times 4x + \frac{\frac{1}{2}(\frac{1}{2} - 1)}{2} \times (4x)^2 + \frac{\frac{1}{2}(\frac{1}{2} - 1)(\frac{1}{2} - 2)}{6} \times (4x)^3 + \ldots$

$= 1 + 2x - 2x^2 + 4x^3 + \ldots$

(b) $|4x| < 1$

$-1 < 4x < 1$

$-\frac{1}{4} < x < \frac{1}{4}$

(c) $x = 0.1$

$\sqrt{1.4} = 1 + 2 \times 0.1 - 2 \times 0.1^2 + 4 \times 0.1^3 + \ldots$

$= 1.18$ (to 3 s.f.)

- Use the formula for the expansion of $(1 + x)^n$, with $n = \frac{1}{2}$ and replacing x by $4x$. Take care when simplifying the coefficients.
- In the formula book it states the expansion is valid for $|x| < 1$. Again replace x by $4x$ and then simplify to obtain the inequality.
- Note that $\sqrt{1.4} = (1 + 4 \times 0.1)^{\frac{1}{2}}$, so the value is obtained by substituting. Be sure to give your answer to the accuracy specified.

Questions to try

Q1 A polynomial $p(x)$ is defined as:
$p(x) = (2 + 3x)^5$

Use the binomial theorem to find the coefficient of x^4 when $p(x)$ is expanded.

Q2 Expand $(x - \frac{1}{x})^3$.

Q3 Expand $(1 + x)^5$, in ascending powers of x, simplifying the coefficients.

Hence by letting $x = y + y^2$, find the coefficient of y^4 in the expansion of $(1 + y + y^2)^5$ in powers of y.

Q4 If $(1 - ax)^n = 1 - 20x + 160x^2 + \ldots$, find the values of a and n.

Q5 **(a)** Expand $(p + qx)^3$.
(b) If $(1 + 2x)(p + qx)^3 = 8 + ax + 30x^2 + bx^3 + \ldots$ where $a > 0$, find the values of p, q, a and b.

Q6 Expand $(1 - 2x)^{-\frac{1}{2}}$, in ascending powers of x, up to and including the term in x^2. State the set of values of x for which the expansion is valid.

Q7 **(a)** Expand $(1 - 3x)^{\frac{1}{3}}$, $|x| < \frac{1}{3}$, in ascending powers of x, up to and including the term in x^3.

(b) By substituting $x = 10^{-3}$ in your expansion, find, to 9 significant figures, the cube root of 997.

Q8 Expand the expression:

$$\frac{1}{1 - x} + \frac{1}{1 - 2x}$$

in ascending powers of x up to and including x^3.

State the range of values of x for which the whole expansion is valid.

Q9 **(a)** Obtain the binomial expansion of $\sqrt{1 + x^2}$, in ascending powers of x, up to and including x^4.
(b) Use your expression to obtain an approximation for:

$$\int_0^{0.4} \sqrt{1 + x^2}\,dx$$

giving your answer correct to 3 decimal places.

Answers can be found on pages 85–86.

2 Trigonometry

Key points to remember

- To prove **trigonometric identities**, begin with one side of the identity and manipulate this expression to get the other. Identities often involve **compound angle formulae**.
  For example, to prove $\sin 2A = 2\sin A\cos A$ start with $\sin 2A$.

  $$\begin{aligned}\sin 2A &= \sin(A + A)\\ &= \sin A\cos A + \sin A\cos A\\ &= 2\sin A\cos A\end{aligned}$$

- To solve equations of the form $a\cos x + b\sin x = c$ first write the left-hand side as $R\cos(x - \alpha)$ or $R\sin(x + \alpha)$ and then find the appropriate values for R and α.

  Using $R\sin(x + \alpha)$ gives $a = R\sin\alpha$ and $b = R\cos\alpha$ so that:

  $\tan\alpha = \dfrac{a}{b}$ and $R^2 = a^2 + b^2$

Formulae you must know

- Compound angle formulae are given in the formula book, but you will benefit from knowing them.

  $\cos(A + B) \equiv \cos A\cos B - \sin A\sin B$ $\quad$ $\cos(A - B) \equiv \cos A\cos B + \sin A\sin B$

  $\sin(A + B) \equiv \sin A\cos B + \cos A\sin A$ $\quad$ $\sin(A - B) \equiv \sin A\cos B - \cos A\sin B$

- Double angle formulae that are not in the formula book.

  $\sin 2A \equiv 2\sin A\cos A$

  $\cos 2A \equiv \cos^2 A - \sin^2 A \equiv 2\cos^2 A - 1 \equiv 1 - 2\sin^2 A$

- Identities that are not in the formula book.

  $\sin^2 A + \cos^2 A \equiv 1$ $\quad$ $1 + \tan^2 A \equiv \sec^2 A$ $\quad$ $1 + \cot^2 A \equiv \mathrm{cosec}^2 A$

- Exact forms for 30°, 45° and 60°.

  $\sin 30° = \cos 60° = \dfrac{1}{2}$ $\quad$ $\sin 60° = \cos 30° = \dfrac{\sqrt{3}}{2}$ $\quad$ $\sin 45° = \cos 45° = \dfrac{1}{\sqrt{2}} = \dfrac{\sqrt{2}}{2}$

  $\tan 30° = \dfrac{1}{\sqrt{3}} = \dfrac{\sqrt{3}}{3}$ $\quad$ $\tan 60° = \sqrt{3}$ $\quad$ $\tan 45° = 1$

Don't make these mistakes ...

Don't give the answers in the **wrong units** (degrees or radians) or in the **wrong range**.

Don't leave out some solutions.

Don't divide by $\sin x$ (or $\cos x$) in a trigonometric equation – remember that $\sin x = 0$ (or $\cos x = 0$) will give solutions to the equation.

Don't use your calculator to simplify when the question asks for answers in **surd form**. Always use exact working when possible.

Exam Questions and Answers

Solve the equation:

$\cot^2\theta + 5\text{cosec}\theta = 3$

giving all solutions in degrees, to the nearest $0.1°$ in the interval $0 \leqslant \theta \leqslant 360°$.

$$\cot^2\theta + 5\text{cosec}\theta = 3$$

$$\text{cosec}^2\theta - 1 + 5\text{cosec}\theta = 3$$

$$\text{cosec}^2\theta + 5\text{cosec}\theta - 4 = 0$$

$$\text{cosec}\theta = \frac{-5 \pm \sqrt{41}}{2}$$

$\text{cosec}\theta = 0.70156$ or -5.70156

Ignore $\text{cosec}\theta = 0.70156$ since $|\text{cosec}\theta| > 1$.

$$\frac{1}{\sin\theta} = -5.70156$$

$$\theta = -10.1°$$

The required solution must be between 0 and 360°.

$\theta = 180° - (-10.1°) = 190.1°$ or
$\theta = 360° - 10.1° = 349.9°$

(a) A student is asked to express $3\sin\theta + 4\cos\theta$ in the form $R\sin(\theta + \alpha)$. She writes:

$3\sin\theta + 4\cos\theta \equiv 5\sin(\theta + \frac{\pi}{3})$

Determine, with a reason, whether she is correct.

(b) Find the general solution, in radians, of the equation $3\sin\theta + 4\cos\theta = 2$.

(a) When $\theta = 0$:

$3\sin\theta + 4\cos\theta = 4$

$5\sin(\theta + \frac{\pi}{3}) = 5\sin\frac{\pi}{3} = \frac{5\sqrt{3}}{2}$

and since these have different values the formulae are not equivalent.

How to score full marks

- This question also contains the additional statement: No credit will be given for a numerical approximation or for a numerical answer without supporting working.
- Make sure you show all your working.
- Use the trigonometric identity $\cot^2\theta + 1 = \text{cosec}^2\theta$ to give a quadratic equation in $\text{cosec}\theta$. Solve the quadratic, using the quadratic equation formula.
- Remember that since $|\sin\theta| < 1$ then $|\text{cosec}\theta| > 1$ so you can reject the solution 0.701 56.
- Make sure that you give all the solutions between 0 and 360°. Use a graph or the CAST method to find the solutions.
- Try some values of θ to find a contradiction: $\theta = 0, \frac{\pi}{6}, \frac{\pi}{3}, \frac{\pi}{2}$ are good values to try. You only need to find one contradiction.

(b) $3\sin\theta + 4\cos\theta \equiv 5\sin(\theta + \alpha)$

where $\tan\alpha = \frac{4}{3}$

$\alpha = 0.9273$

$5\sin(\theta + 0.9273) = 2$

$\theta + 0.9273 = \sin^{-1}\frac{2}{5}$

$\theta = 1.80$ or $\theta = 5.77$

General solutions: $\theta = 1.80 + 2n\pi$ or $\theta = 5.77 + 2n\pi$

How to score full marks

- We are comparing $5\sin\theta\cos\alpha + 5\cos\theta\sin\alpha$ with $3\sin\alpha + 4\cos\alpha$. So $\sin\alpha = \frac{4}{5}$, $\cos\alpha = \frac{3}{5}$ giving $\tan\alpha = \frac{4}{3}$.
- Remember to use radians as stated in the question. These solutions are in the range $0 \leqslant \theta \leqslant 2\pi$. You also need to state the general solutions.

Q3 **(a)** Prove that $\cos 3\theta \equiv 4\cos^3\theta - 3\cos\theta$.

(b) Hence find $\int_0^{\pi} \cos^3\theta d\theta$.

(a)
$$\begin{aligned}\cos 3\theta &= \cos(2\theta + \theta)\\ &= \cos 2\theta\cos\theta - \sin 2\theta\sin\theta\\ &= (2\cos^2\theta - 1)\cos\theta - 2\sin\theta\cos\theta\sin\theta\\ &= 2\cos^3\theta - \cos\theta - 2\sin^2\theta\cos\theta\\ &= 2\cos^3\theta - \cos\theta - 2(1 - \cos^2\theta)\cos\theta\\ &= 4\cos^3\theta - 3\cos\theta\end{aligned}$$

(b) $\cos^3\theta = \frac{1}{4}\cos 3\theta + \frac{3}{4}\cos\theta$

$$\int_0^{\frac{\pi}{2}} \cos^3\theta d\theta = \int_0^{\frac{\pi}{2}} (\tfrac{1}{4}\cos 3\theta + \tfrac{3}{4}\cos\theta)d\theta$$

$$= \left[\tfrac{1}{12}\sin 3\theta + \tfrac{3}{4}\sin\theta\right]_0^{\frac{\pi}{2}}$$

$$= (-\tfrac{1}{12} + \tfrac{3}{4}) - 0$$

$$= \tfrac{2}{3}$$

- Start with the left-hand side of the identity. First use the addition formula and then substitute, using $\cos 2\theta = 2\cos^2\theta - 1$ and $\sin 2\theta = 2\sin\theta\cos\theta$. Replace the $\sin^2\theta$ by $1 - \cos^2\theta$ and simplify to prove the result.
- Rearrange the above expression to make $\cos^3\theta$ the subject. Integrate using: $\int \cos kx dx = \frac{1}{k}\sin kx$.
- Substitute the limits of integration and simplify to obtain the result.

Questions to try

Q1 Prove the identity $\sin(x + 30°) + \sqrt{3}\cos(x + 30°) \equiv 2\cos x$ where x is measured in degrees.

Hence express $\cos 15°$ in surd form.

Q2 **(a)** Prove the identity $\tan x + \cot x \equiv \dfrac{2}{\sin 2x}$.

(b) Given that $0 < x < \dfrac{\pi}{4}$, find the values of x for which $\tan x + \cot x > 4$.

(c) The region enclosed by the x-axis, the curve $y = \tan x + \cot x$ and the lines $x = \dfrac{\pi}{8}$ and $x = \dfrac{\pi}{6}$ is denoted by R. Show that the volume generated when R is rotated through 360° about the x-axis is $\dfrac{2\pi}{3}(3 - \sqrt{3})$.

Q3 **(a)** Prove that $\sin 4x \equiv 4\sin x\cos x\cos 2x$.
(b) Hence, without using a calculator, find the value of $\sin 4x$ when $\sin x = \frac{3}{5}$.

Q4 **(a)** Find the positive constant R and the acute angle A for which:
$\cos x + \sin x \equiv R\cos(x - A)$
(b) Find the general solution, in radians, of the equation $\cos x + \sin x = 1$.
(c) Deduce the greatest value of $\cos x + \sin x$.

Q5 Prove that $\sin 3\theta = 3\sin\theta - 4\sin^3\theta$

Hence find all values of θ for $0 < \theta < 360°$ which satisfy the equation $\sin 3\theta = 2\sin\theta$.

Q6 The function f is defined by $f(x) = \sin 2x + 2\cos^2 x$ where x is real and measured in radians.
(a) Find the general solution of the equation $f(x) = 0$.
(b) Prove that $1 - \sqrt{2} \leqslant f(x) \leqslant 1 + \sqrt{2}$.
(c) The diagram shows part of the curve with equation $y = f(x)$.
Find the area of the shaded region, giving your answer in terms of π.

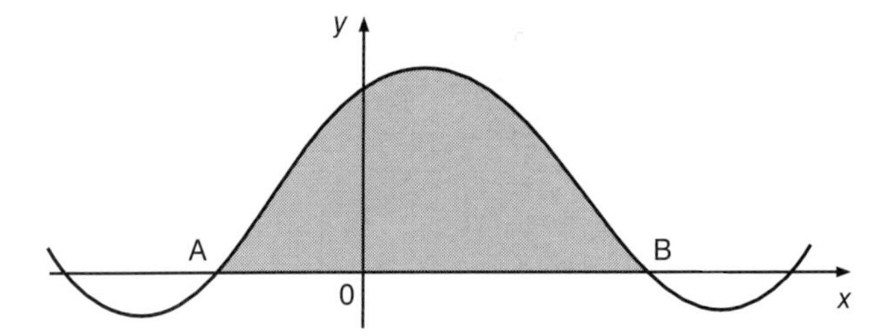

Q7 The diagram shows the rectangular cross-section PQRS of a letter rack. A rectangular envelope ABCD rests in the vertical plane PQRS inside the letter rack. QR is horizontal. QR = 30 cm, AD = 27 cm and CD = 18 cm. The bottom edge, BC, of the envelope, makes an angle $x°$ with the base QR of the rack.
(a) Prove that $9\cos x° + 6\sin x° = 10$.
(b) Express $9\cos x° + 6\sin x°$ in the form $R\cos(x° - \alpha°)$, where $R > 0$ and $0 < \alpha < 90$, giving the values of R and α to 2 decimal places.
(c) Hence, or otherwise, find x, giving your answer to the nearest tenth of a degree.

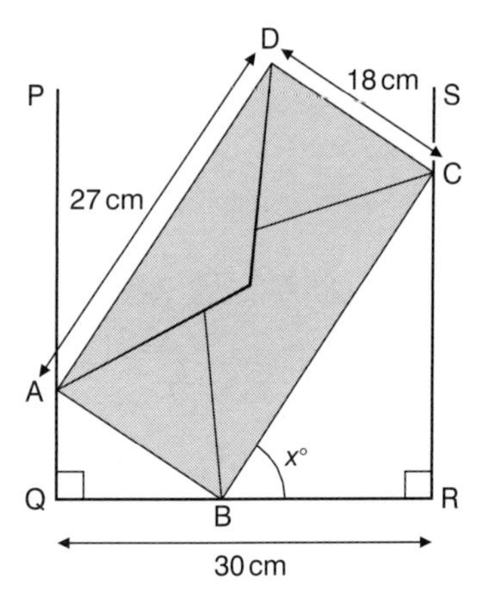

Answers can be found on pages 86–88.

3 Differentiation techniques

Key points to remember

- Use the **chain rule** to differentiate expressions such as $y = \sin x^3$, writing $y = \sin u$ and $u = x^3$.
- Use the **product rule** for expressions such as $y = x^3 \sin 2x$, writing $u = x^3$ and $v = \sin 2x$.
- Use the **quotient rule** for expressions such as $y = \frac{e^x}{\cos x}$, writing $u = e^x$ and $v = \cos x$.
- Use **parametric differentiation** when x and y are expressed in terms of a **parameter** such as t or θ.
  For example, if $x = \sin t$ and $y = \cos t$, then $\frac{dx}{dt} = \cos t$ and $\frac{dy}{dt} = -\sin t$.

  Using the formula then gives:
  $\frac{dy}{dx} = -\frac{\sin t}{\cos t} = -\tan t$
- Differentiate **exponential functions** such as a^x with respect to x by first taking logarithms.
- Use **implicit differentiation** for expressions that cannot easily be written in the form $y = f(x)$.
  Differentiate term by term and make $\frac{dy}{dx}$ the subject.
- **Tangents to curves** have the equation $y = mx + c$, where m is found by evaluating $\frac{dy}{dx}$ at the point where the tangent touches the curve.
- **Normals to curves** have the equation $y = mx + c$, where m is found by evaluating $\frac{-1}{\frac{dy}{dx}}$ at the point where the normal crosses the curve.

Fomulae you need to know

- $\frac{dy}{dx} = \frac{dy}{du} \times \frac{du}{dx}$ (chain rule)
- $\frac{d}{dx}(uv) = v\frac{du}{dx} + u\frac{dv}{dx}$ (product rule)
- $\frac{d}{dx}\left(\frac{u}{v}\right) = \frac{v\frac{du}{dx} - u\frac{dv}{dx}}{v^2}$ (quotient rule)
- $\frac{dy}{dx} = \frac{\frac{dy}{dt}}{\frac{dx}{dt}}$ (parametric differentiation)
- $\frac{d}{dx}(x^n) = nx^{n-1}$
- $\frac{d}{dx}(\sin kx) = k\cos kx$
- $\frac{d}{dx}(\cos kx) = -k\sin kx$
- $\frac{d}{dx}(\tan kx) = k\sec^2 kx$
- $\frac{d}{dx}(e^{kx}) = ke^{kx}$
- $\frac{d}{dx}(\ln x) = \frac{1}{x}$

Don't make these mistakes...

Don't use incorrect versions of the **product**, **quotient** or **chain** rules.

Don't forget to use the **product rule** for expressions like xy when you are doing **implicit differentiation**.

Don't make **algebraic errors** when simplifying expressions.

Exam Questions and Answers

Q1 **(a)** Differentiate $y = \frac{\cos 2x}{x^3}$ given that $x > 0$.

(b) The curve has a stationary point when $x = a$.

Show that $\frac{3}{2a} = -\tan 2a$.

(a) $u = \cos 2x$ and $v = x^3$

$\frac{du}{dx} = -2\sin 2x$ $\quad \frac{du}{dx} = 3x^2$

$$\frac{d}{dx}\left(\frac{u}{v}\right) = \frac{v\frac{du}{dx} - u\frac{dv}{dx}}{v^2}$$

$$= \frac{x^3(-2\sin 2x) - \cos 2x \times 3x^2}{x^6}$$

$$= \frac{-(2x\sin 2x + 3\cos 2x)}{x^4}$$

(b) $\frac{-(2a\sin 2a + 3\cos 2a)}{a^4} = 0$

$2a\sin 2a + 3\cos 2a = 0$

$3\cos 2a = -2a\sin 2a$

$\frac{3}{2a} = -\tan 2a$

Q2 Given that $y = 4^x$, show that $\frac{dy}{dx} = 4^x \ln 4$.

$\ln y = \ln 4^x$

$\ln y = x\ln 4$

$\frac{1}{y} \times \frac{dy}{dx} = \ln 4$

$\frac{dy}{dx} = y\ln 4 = 4^x \ln 4$

How to score full marks

- You will need to use the quotient rule to differentiate this expression.
- You need to know the formula for differentiating $\cos kx$.
- Apply the rule carefully and don't leave out the minus signs.
- Simplify the result, noting that x^2 is a factor of both terms on the top.
- As there is a stationary point at $x = a$, substitute a for x and set $\frac{dy}{dx} = 0$.
- Rearrange into the required form, using $\frac{\sin\theta}{\cos\theta} = \tan\theta$.
- Make sure that you show all the steps that you need to take, so that you satisfy the examiner.
- First take logs of both sides and simplify. Then differentiate both sides, using implicit differentiation.
- Finally substitute 4^x for y to reach the required result.

How to score full marks

A curve has equation $x^2 + xy + y^3 = 7$. Find:

(a) $\frac{dy}{dx}$

(b) the equation of the tangent to the curve at the point (2, 1).

(a) $2x + (x \times \frac{dy}{dx} + 1 \times y) + 3y^2 \times \frac{dy}{dx} = 0$

$$2x + x\frac{dy}{dx} + y + 3y^2\frac{dy}{dx} = 0$$

$$2x + y + \frac{dy}{dx}(x + 3y^2) = 0$$

$$\frac{dy}{dx} = -\frac{2x + y}{x + 3y^2}$$

- You need to use implicit differentiation for this expression. The term xy is a product, so apply the product rule to give the derivative shown in the bracket.
- After differentiating, you need to make $\frac{dy}{dx}$ the subject.

(b) $\frac{dy}{dx} = -\frac{4 + 1}{2 + 3} = -1$

$y = -x + c$

$1 = -2 + c \Rightarrow c = 3$

$y = 3 - x$

- Use $x = 2$ and $y = 1$ to find the gradient of the curve at this point. This gives the gradient of the tangent. Then use the equation of a straight line for the tangent to find the value of c, substituting for the point given.

A curve is defined parametrically as $x = 2\cos t$, $y = 3\sin t$.

Find the equation of the normal to the curve when $t = \frac{\pi}{4}$, giving your answer in exact form.

$x = 2\cos t$ so $\frac{dx}{dt} = -2\sin t$

$y = 3\sin t$ so $\frac{dy}{dt} = 3\cos t$

$\frac{dy}{dx} = \frac{3\cos t}{-2\sin t} = -\frac{3}{2\tan t}$

- You need to use parametric differentiation here, so find $\frac{dy}{dt}$ and $\frac{dx}{dt}$ first, then substitute them into $\frac{dy}{dx} = \frac{\frac{dy}{dt}}{\frac{dx}{dt}}$.

$t = \frac{\pi}{4} \Rightarrow \frac{dy}{dx} = -\frac{3}{2}$

When $t = \frac{\pi}{4}$, $x = 2\cos\frac{\pi}{4} = \sqrt{2}$

and $y = 3\sin\frac{\pi}{4} = \frac{3}{\sqrt{2}}$.

- Substituting $t = \frac{\pi}{4}$ into $\frac{dy}{dx}$ gives the gradient of the tangent at this point. The coordinates of the point can also be found in exact form.

$y = \frac{2}{3}x + c$

$\frac{3}{\sqrt{2}} = \frac{2}{3} \times \sqrt{2} + c$

$c = \frac{3}{\sqrt{2}} - \frac{2\sqrt{2}}{3} = \frac{9 - 2\sqrt{2}\sqrt{2}}{3\sqrt{2}} = \frac{9 - 4}{3\sqrt{2}} = \frac{5}{3\sqrt{2}}$

$y = \frac{2}{3}x + \frac{5}{3\sqrt{2}}$

- The gradient of the normal will be $\frac{2}{3}$, so you can use this to reach an equation of the form $y = mx + c$. Then use the coordinates of the point to find the value of c and hence the equation.

Questions to try

Q1 The equation of a curve is $y\cos x = x + y^2$. Find the gradient of the curve at the point (0, 1).

Q2 The volume $V\,\text{cm}^3$ in a container, when the depth is $x\,\text{cm}$ $(x > 0)$, is given by:

$$V = \frac{30\sqrt{x}}{x+9}$$

The container has height $h\,\text{cm}$.

Given that $x = h$ when $\frac{dV}{dx} = 0$, find the value of h, and hence determine the value of V when $x = h$.

Q3 Find the equation of the normal to the curve $y = \frac{2x+3}{6-x}$ at the point (5, 13).

Q4 A curve has equation $y = (x-4)e^{-x}$. Find the coordinates of the stationary point on this curve and determine the nature of the stationary point.

Q5 A curve is defined parametrically by:

$x = 2t^2 - 1,\ y = t^3$

and P is the point on the curve where $t = 2$.

Obtain an expression for $\frac{dy}{dx}$ in terms of t and calculate the gradient of the curve at P.

Q6 A curve is given by the parametric equations:

$x = 4\sin^3 t,\ y = \cos 2t,\ 0 \leqslant t \leqslant \frac{\pi}{4}$

(a) Show that $\frac{dx}{dy} = -3\sin t$.

(b) Find an equation of the normal to the curve at the point where $t = \frac{\pi}{6}$.

Q7 The parametric equations of a curve are:

$x = a\sin\theta,\ y = a\theta\cos\theta$

where a is a positive constant and $0 < \theta < \frac{\pi}{2}$. Find $\frac{dy}{dx}$ in terms of θ, and hence show that the gradient of the curve is zero where $\tan\theta = \frac{1}{\theta}$.

By sketching a suitable pair of graphs, show that the equation $\tan\theta = \frac{1}{\theta}$ is satisfied by just one value of θ in the relevant range.

Determine, with reasons, whether this value of θ is greater or less than $\frac{\pi}{4}$.

Q8 Find $\frac{dy}{dx}$ if $y = \frac{x}{\sqrt{x^2+1}}$ and hence find $\int \frac{2}{(x^2+1)^{\frac{3}{2}}}\,dx$.

Answers can be found on pages 88–90.

4 Integration

Key points to remember

- You must know the **standard integrals** listed in the formulae below.
- Use integration to find the **area enclosed by a curve**.

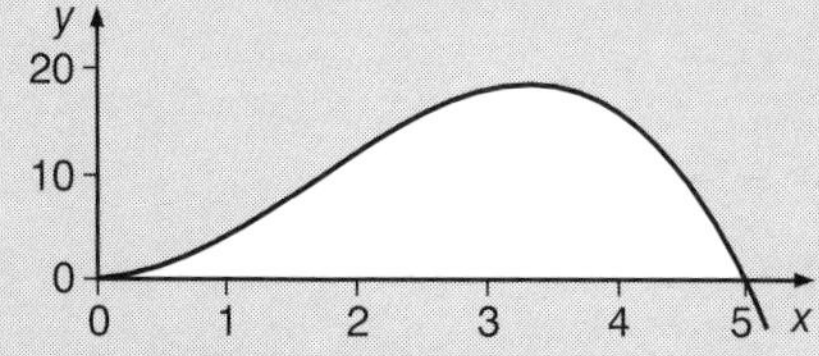

This area is given by:

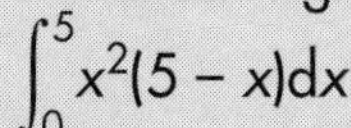

$$\int_0^5 x^2(5-x)\,dx$$

- Use the method of **integration by parts** for expressions such as:

$$\int x\sin x\,dx$$

In this example, take $u = x$ and $\frac{dy}{dx} = \sin x$, then use the formula:

$$\int u\frac{dv}{dx}dx = uv - \int v\frac{du}{dx}dx$$

which will be included in your formula book. You do not need to learn it, but you must be able to find it and know how to use it. Always choose u to simplify the product that you need to integrate.

- Use the method of **integration by substitution** for problems such as:

$$\int x^2\sin x^3\,dx$$

In this example, take $u = x^3$. For expressions such as $\int \frac{1}{\sqrt{1-x^2}}dx$ use a substitution such as $x = \sin\theta$.

- Calculate a **volume of revolution** (about the x-axis) by using the formula:

$$V = \pi\int y^2\,dx$$

For example, the volume formed by rotating the region enclosed by the curve $y = 4x - x^2$ and the x-axis is given by $V = \pi\int(4x - x^2)^2dx$.

Formulae you must know

- $\int x^n dx = \frac{x^{n+1}}{n+1} + c,\ n \neq -1$
- $\int \cos kx\,dx = \frac{1}{k}\sin kx + c$
- $\int \sin kx\,dx = -\frac{1}{k}\cos kx + c$
- $\int \sec^2 kx\,dx = \frac{1}{k}\tan kx + c$
- $\int e^{kx}dx = \frac{1}{k}e^{kx} + c$
- $\int \frac{1}{x}dx = \ln|x| + c$
- $V = \pi\int y^2dx$

Don't make these mistakes...

If a substitution you have chosen does not seem to work, **stop and try another**. Don't waste time going nowhere. The same applies to integration by parts.

Don't fail to **use a hint** that is given in the first part of a question. Examiners may ask you to prove a result in one part that will help you in an integration in the second part of a question.

Don't forget to **include a constant of integration**.

Don't forget to **change the variables back** at the end of an integration by substitution.

Don't miss out terms when squaring an expression to find a volume of revolution.

Exam Questions and Answers

Q1 Find $\int xe^{3x}dx$.

$u = x$ and $\frac{dv}{dx} = e^{3x}$

$\frac{du}{dx} = 1$ $v = \frac{1}{3}e^{3x}$

Then the integral becomes:

$$\int u\frac{dv}{dx}dx = uv - \int v\frac{du}{dx}dx$$

$$\int xe^{3x}dx = \frac{xe^{3x}}{3} - \int \frac{1}{3}e^{3x}dx$$

$$= \frac{xe^{3x}}{3} - \frac{e^{3x}}{9} + c$$

$$= \frac{e^{3x}}{9}(3x - 1) + c$$

Q2 Find $\int \sin^4 x\cos x dx$.

$u = \sin x$

$\frac{du}{dx} = \cos x$ and $\frac{dx}{du} = \frac{1}{\cos x}$

$$\int \sin^4 x\cos x dx = \int u^4 \cos x \frac{1}{\cos x}du$$

$$= \int u^4 du$$

$$= \frac{u^5}{5} + c$$

$$= \frac{1}{5}\sin^5 x + c$$

How to score full marks

- Use integration by parts. Set out your working clearly so that you don't make mistakes. In this case differentiate the x-term so that the new integral only contains exponential terms.
- This will be in your formula book. Substitute carefully and then integrate to obtain the final result.
- Don't forget the constant of integration.
- You need to use integration by substitution. You could try $u = \sin x$ or $u = \cos x$. The first of these will simplify the integral but the second will make it more difficult. Note how the terms in $\cos x$ cancel out once the correct substitution has been made.
- Don't forget to give the result in terms of x and to include a constant of integration.

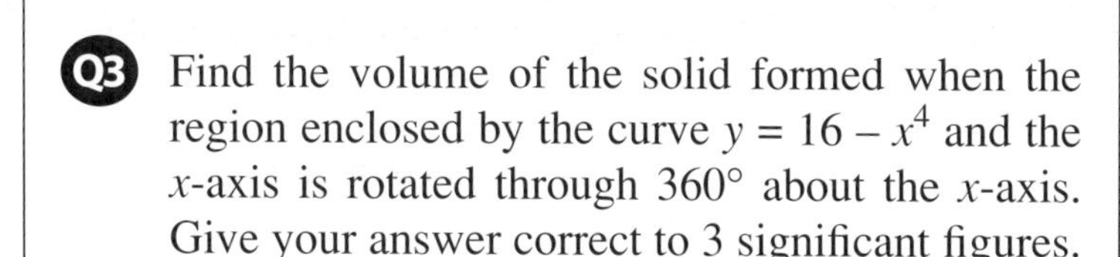

Q3 Find the volume of the solid formed when the region enclosed by the curve $y = 16 - x^4$ and the x-axis is rotated through 360° about the x-axis. Give your answer correct to 3 significant figures.

The curve intersects the x-axis when y = 0.

$16 - x^4 = 0$

$x^4 = 16$

$x = -2$ or 2

$$V = \pi \int_{-2}^{2} (16 - x^4)^2 dx$$

$$= \pi \int_{-2}^{2} (x^8 - 32x^4 + 256) dx$$

$$= \pi \left[\frac{x^9}{9} - \frac{32x^5}{5} + 256x \right]_{-2}^{2}$$

= 2290 cubic units (3 s.f.)

Q4 Find $\int_0^{\pi} \sin^2 \frac{x}{2} dx$.

$\cos x = 1 - 2\sin^2 \frac{x}{2}$

$$\text{so } \int_0^{\pi} \sin^2 \frac{x}{2} dx = \int_0^{\pi} \left(\frac{1}{2} - \frac{1}{2}\cos x\right) dx$$

$$= \left[\frac{1}{2}x - \frac{1}{2}\sin x \right]_0^{\pi}$$

$$= \frac{\pi}{2}$$

How to score full marks

- First you need to find where the curve intersects the x-axis.
- Then use the formula $V = \pi \int y^2 dx$ to find the volume. In problems like this don't make silly mistakes when expanding the brackets.
- Give the answer to 3 significant figures as specified.
- You need to start by using the double angle formula $\cos 2A = 1 - 2\sin^2 A$. In this case $A = \frac{x}{2}$, so $2A = x$.

  In an exam question you may be asked to show this first, using the formula for $\cos(A + B)$ which is given in your formula book.
- Note that $\sin 0 = 0$ and $\sin \pi = 0$.

Questions to try

Q1 The finite region bounded by the curve with equation $y = x - x^2$ and the x-axis is rotated through 360° about the x-axis. Using integration, find, in terms of π, the volume of its solid form.

Q2 The diagram shows the region R which is bounded by the curve $y = \frac{2}{x+1}$, the x-axis, and the lines $x = 1$ and $x = 5$.

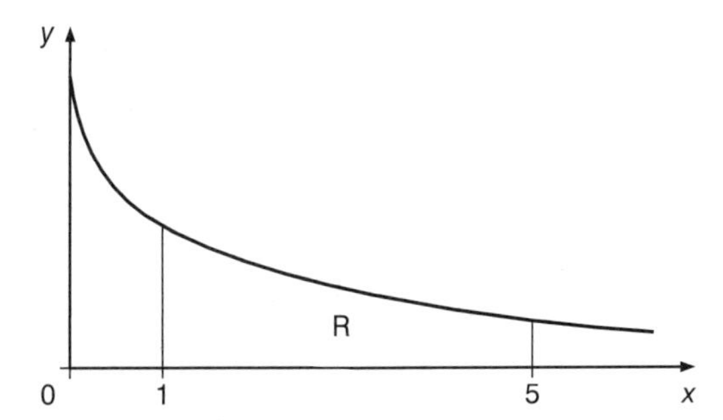

Use integration:

(a) to find the area of R, giving your answer as a single logarithm

(b) to show that the volume of the solid formed when R is rotated completely about the x-axis is $\frac{4\pi}{3}$.

Q3 By using the substitution $u = \sin x$, or otherwise, find:

$$\int \sin^3 x \sin 2x \, dx$$

giving your answer in terms of x.

Q4 A curve is defined for $x > 0$ and has equation $y = x^2 \ln 4x$. Show that the area bounded by the curve, the x-axis and the lines $x = 1$ and $x = 2$ is $\frac{66 \ln 2 - 7}{9}$.

Q5 Find $\int x^2 \cos x^3 \, dx$.

Q6 Given that $a > 1$, find $\int_1^a x \ln x \, dx$ in terms of a.

Q7 **(a)** Use the identities for $\cos(A + B)$ and $\cos(A - B)$ to prove that:

(i) $2\cos A \cos B \equiv \cos(A + B) + \cos(A - B)$

(ii) $\cos^2 A = \frac{1}{2}(1 + \cos 2A)$.

(b) Find $\int \cos 3x \cos x \, dx$.

(c) Use the substitution $x = \cos t$ to evaluate $\int_0^{\frac{1}{2}} \frac{x^2}{(1 - x^2)^{\frac{1}{2}}} \, dx$.

Answers can be found on pages 90–91.

Key points to remember

- A **vector** quantity has **magnitude** and **direction**.
- A **scalar** quantity only has **magnitude**.
- In two dimensions, you write a vector in **component** form in terms of perpendicular unit vectors **i** (→) and **j**(↑) as:

  $\mathbf{a} = \begin{pmatrix} a_1 \\ a_2 \end{pmatrix}$ or $\mathbf{a} = a_1\mathbf{i} + a_2\mathbf{j}$
- The position vector of a point P is the vector $\overrightarrow{OP}$ from the origin to the point.
- Vectors are added according to the **triangle rule of vector addition**.
- In three dimensions, you write a vector in component form as:

  $\mathbf{a} = \begin{pmatrix} a_1 \\ a_2 \\ a_3 \end{pmatrix}$ or $\mathbf{a} = a_1\mathbf{i} + a_2\mathbf{j} + a_3\mathbf{k}$

Don't confuse vectors and coordinates.

Don't subtract vectors in the wrong order.

Don't forget, to find $\overrightarrow{AB}$ you need **b** – **a**, not **a** – **b**.

Formulae you must know

- The **magnitude** of **a** (where $\mathbf{a} = a_1\mathbf{i} + a_2\mathbf{j}$) is $|\mathbf{a}| = \ a_1^2 + a_2^2$.
- The **direction** of **a** is the angle **a** makes with the *x*-axis measured in an anti-clockwise direction.

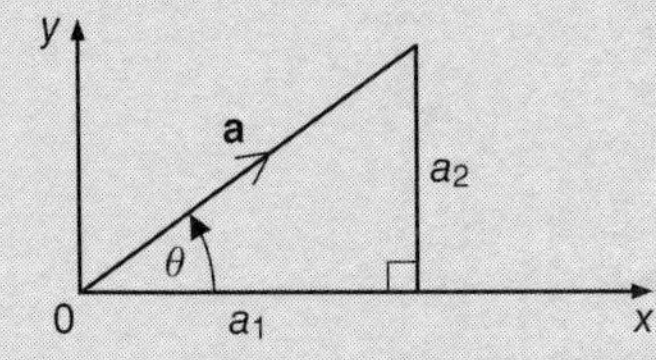

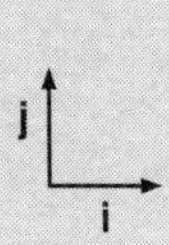

- The magnitude of **a** (where $\mathbf{a} = a_1\mathbf{i} + a_2\mathbf{j} + a_3\mathbf{k}$) is $|\mathbf{a}| = \sqrt{a_1^2 + a_2^2 + a_3^2}$.
- The **scalar product** of two vectors **a** and **b**, acting in directions at angle θ to each other, is given by:

  $\mathbf{a}.\mathbf{b} = |\mathbf{a}||\mathbf{b}|\cos\theta$
  $= a_1b_1 + a_2b_2$ in two dimensions
  $= a_1b_1 + a_2b_2 + a_3b_3$ in three dimensions

  Using column vectors:

  $$\mathbf{a}.\mathbf{b} = \begin{pmatrix} a_1 \\ a_2 \\ a_3 \end{pmatrix} . \begin{pmatrix} b_1 \\ b_2 \\ b_3 \end{pmatrix} = a_1b_1 + a_2b_2 + a_3b_3$$
- If two vectors are perpendicular, then $\theta = 90°$ and $\mathbf{a}.\mathbf{b} = 0$.
- The vector equation of a line through points A and B with positions vectors **a** and **b** is:

  $\mathbf{r} = \mathbf{a} + t(\mathbf{b} - \mathbf{a}) = (1 - t)\mathbf{a} + t\mathbf{b}$
- The vector equation of a line through point A with position vector **a** in the direction of a vector **d** is: $\mathbf{r} = \mathbf{a} + t\mathbf{d}$.

Exam Questions and Answers

Q1 Three points P, Q and R have position vectors **p**, **q** and **r** respectively, where $\mathbf{p} = 7\mathbf{i} + 10\mathbf{j}$, $\mathbf{q} = 3\mathbf{i} + 12\mathbf{j}$, $\mathbf{r} = -\mathbf{i} + 4\mathbf{j}$.

(a) Write down the vectors $\overrightarrow{PQ}$ and $\overrightarrow{RQ}$ and show that they are perpendicular.

(b) Using a scalar product, or otherwise, find the angle PRQ.

(c) Find the position vector of S, the midpoint of PR.

(d) Show that $|\overrightarrow{QS}| = |\overrightarrow{RS}|$.

Using your previous results, or otherwise, find the angle PSQ.

(a) $\overrightarrow{PQ} = \underset{\sim}{q} - \underset{\sim}{p} = (3\underset{\sim}{i} + 12\underset{\sim}{j}) - (7\underset{\sim}{i} + 10\underset{\sim}{j})$

$= -4\underset{\sim}{i} + 2\underset{\sim}{j}$

$\overrightarrow{RQ} = \underset{\sim}{q} - \underset{\sim}{r} = (3\underset{\sim}{i} + 12\underset{\sim}{j}) - (-\underset{\sim}{i} + 4\underset{\sim}{j})$

$= 4\underset{\sim}{i} + 8\underset{\sim}{j}$

$\overrightarrow{PQ}.\overrightarrow{QR} = (-4\underset{\sim}{i} + 2\underset{\sim}{j}).(4\underset{\sim}{i} + 8\underset{\sim}{j})$

$= (-4) \times (4) + (2) \times (8) = 0$

So $\overrightarrow{PQ}$ and $\overrightarrow{RQ}$ are perpendicular.

- Remember to use the triangle law when adding vectors.

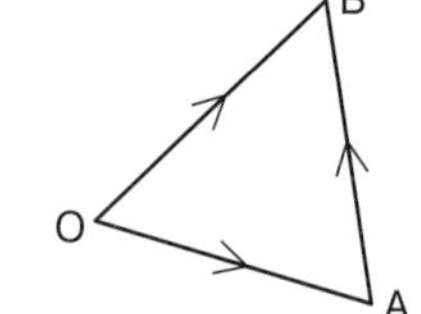

$\overrightarrow{OB} = \overrightarrow{OA} + \overrightarrow{AB}$

$\overrightarrow{AB} = \overrightarrow{OB} - \overrightarrow{OA}$

- If two vectors are perpendicular then their scalar product (dot product) is zero.

(b)

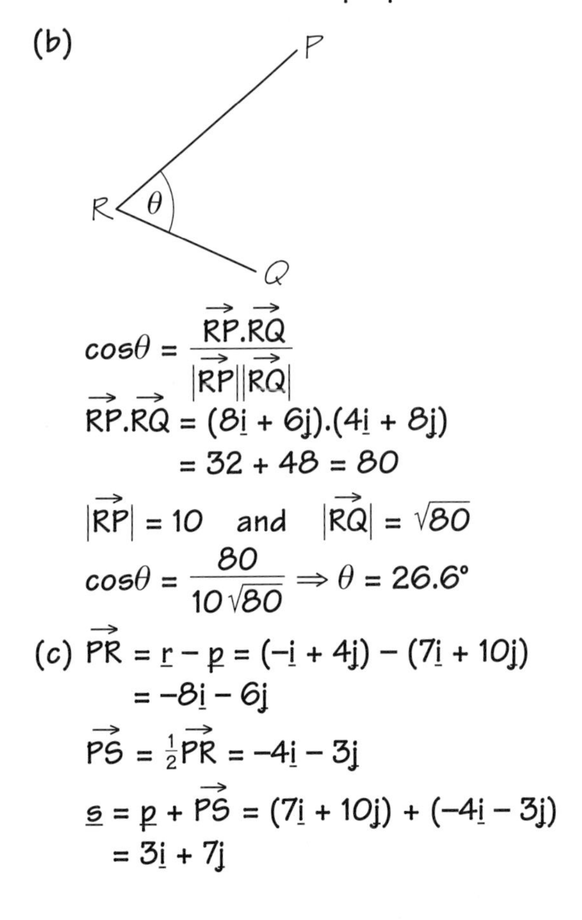

$\cos\theta = \dfrac{\overrightarrow{RP}.\overrightarrow{RQ}}{|\overrightarrow{RP}||\overrightarrow{RQ}|}$

$\overrightarrow{RP}.\overrightarrow{RQ} = (8\underset{\sim}{i} + 6\underset{\sim}{j}).(4\underset{\sim}{i} + 8\underset{\sim}{j})$

$= 32 + 48 = 80$

$|\overrightarrow{RP}| = 10$ and $|\overrightarrow{RQ}| = \sqrt{80}$

$\cos\theta = \dfrac{80}{10\sqrt{80}} \Rightarrow \theta = 26.6°$

- Use the definition of the scalar product to find $\cos\theta$ where θ is the angle between the vectors.

(c) $\overrightarrow{PR} = \underset{\sim}{r} - \underset{\sim}{p} = (-\underset{\sim}{i} + 4\underset{\sim}{j}) - (7\underset{\sim}{i} + 10\underset{\sim}{j})$

$= -8\underset{\sim}{i} - 6\underset{\sim}{j}$

$\overrightarrow{PS} = \frac{1}{2}\overrightarrow{PR} = -4\underset{\sim}{i} - 3\underset{\sim}{j}$

$\underset{\sim}{s} = \underset{\sim}{p} + \overrightarrow{PS} = (7\underset{\sim}{i} + 10\underset{\sim}{j}) + (-4\underset{\sim}{i} - 3\underset{\sim}{j})$

$= 3\underset{\sim}{i} + 7\underset{\sim}{j}$

How to score full marks

How to score full marks

(d) $|\overrightarrow{QS}| = |\underline{s} - \underline{q}| = |-5\underline{j}| = 5$

$|\overrightarrow{RS}| = |\underline{s} - \underline{r}| = |4\underline{i} + 3\underline{j}| = 5$

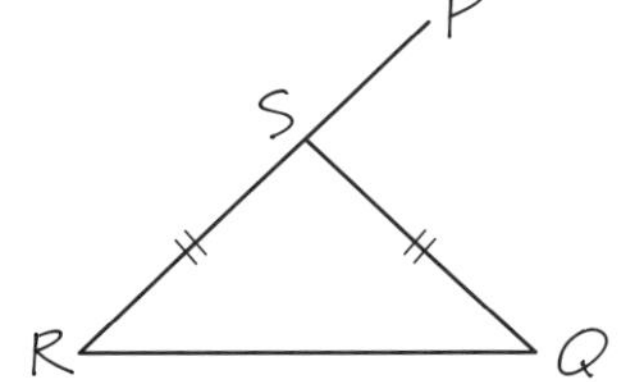

Triangle SRQ is isosceles.

$\angle SRQ = \angle SQR = 26.6°$

$\angle PSQ = 2\angle SRQ = 53.2°$

- Use the angle found in part (b). From the figure: $\angle PSQ = 180° - \angle RSQ$ and $\angle SRQ = 180° - 2 \times \angle SRQ$

Q2 The points A and B have position vectors $\mathbf{a} = 4\mathbf{i} + 5\mathbf{j} + 6\mathbf{k}$ and $\mathbf{b} = 4\mathbf{i} + 6\mathbf{j} + 2\mathbf{k}$ respectively, relative to a fixed origin O. The line l_1 has vector equation $\mathbf{r} = \mathbf{i} + 5\mathbf{j} - 3\mathbf{k} + s(\mathbf{i} + \mathbf{j} - \mathbf{k})$ where s is a scalar parameter.

(a) Write down a vector equation for the line l_2 which passes though the points A and B, giving the equation in terms of a scalar parameter t.

(b) Show that the lines l_1 and l_2 intersect and state the position vector of the point of intersection.

(c) Calculate the acute angle between the lines l_1 and l_2, giving your answer to the nearest tenth of a degree.

(a) $r = \overrightarrow{OA} + t(\overrightarrow{AB})$

$= (4\underline{i} + 5\underline{j} + 6\underline{k}) + t(\underline{j} - 4\underline{k})$

- AB = **b** – **a**

(b) $\underline{i} + 5\underline{j} - 3\underline{k} + s(\underline{i} + \underline{j} - \underline{k})$

$= (4\underline{i} + 5\underline{j} + 6\underline{k}) + t(\underline{j} - 4\underline{k})$

$(s - 3)\underline{i} + (s - t)\underline{j} + (-9 - s + 4t)\underline{k} = 0$

$s = 3,\ t = s = 3$

Check: $-9 - s + 4t = -9 - 3 + 12 = 0$

The point of intersection is $4\underline{i} + 8\underline{j} - 6\underline{k}$.

- Put the vector equations equal to each other and solve for the parameters s and t. Since you have three equations you can use two to solve for s and t and the third to check that your solution is correct.

(c) The direction of l_1 is $\underline{i} + \underline{j} - \underline{k}$

The direction of l_2 is $\underline{j} - 4\underline{k}$

The angle θ between the lines is given by:

$$\cos\theta = \frac{(\underline{i} + \underline{j} - \underline{k}).(\underline{j} - 4\underline{k})}{\sqrt{3}\sqrt{17}} = \frac{5}{\sqrt{51}}$$

$\theta = 45.6°$

- Use the scalar product to find the angle between the two vectors.

Questions to try

Q1 The position vectors of two points A and B on a line are:
$\mathbf{a} = 4\mathbf{i} + 2\mathbf{j} - \mathbf{k}$
$\mathbf{b} = -2\mathbf{i} + 26\mathbf{j} + 11\mathbf{k}$

(a) Show that the vector $2\mathbf{i} - 3\mathbf{j} + 7\mathbf{k}$ is perpendicular to $\overrightarrow{AB}$.
(b) Find the equation of the line.

Q2 The points A and B have coordinates (3, 2, 4) and (4, 4, –3) respectively. The line l_1, which passes through A, has equation:

$$\mathbf{r} = \begin{pmatrix} 3 \\ 2 \\ 4 \end{pmatrix} + t\begin{pmatrix} 5 \\ 1 \\ 1 \end{pmatrix}$$

Show that AB is perpendicular to l_1.

The line l_2, which passes through B, has equation:

$$\mathbf{r} = \begin{pmatrix} 4 \\ 4 \\ -3 \end{pmatrix} + s\begin{pmatrix} 2 \\ 1 \\ -2 \end{pmatrix}$$

Show that the lines l_1 and l_2 intersect and find the coordinates of their point of intersection.

Q3 Referred to a fixed origin O, the points A and B have position vectors $3\mathbf{i} - \mathbf{j} + 2\mathbf{k}$ and $-\mathbf{i} + \mathbf{j} + 9\mathbf{k}$ respectively.

(a) Show that OA is perpendicular to AB.
(b) Find, in vector form, an equation of the line L_1 which passes through A and B.
The line L_2 has equation $\mathbf{r} = 8\mathbf{i} + \mathbf{j} - 6\mathbf{k} + \mu(\mathbf{i} - 2\mathbf{j} - 2\mathbf{k})$, where μ is a scalar parameter.
(c) Show that the lines L_1 and L_2 intersect and find the position vector of their point of intersection.
(d) Calculate, to the nearest tenth of a degree, the acute angle between L_1 and L_2.

Q4 The lines P and Q have vector equations:

$$P\text{: } \mathbf{r} = \begin{pmatrix} 2 \\ 1 \\ 3 \end{pmatrix} + s\begin{pmatrix} 1 \\ 3 \\ -5 \end{pmatrix} \quad \text{and}$$

$$Q\text{: } \mathbf{r} = \begin{pmatrix} -4 \\ 3 \\ 5 \end{pmatrix} + t\begin{pmatrix} 1 \\ -2 \\ 2 \end{pmatrix}$$

(a) Show that the lines P and Q meet and find the coordinates of their point of intersection.
(b) Find the angle between the two lines.

Answers can be found on pages 91–92.

6 Proof

Key points to remember

- **The language of proof**
  - $P \Rightarrow Q$ If statement P is true then so is statement Q. P is a sufficient condition for Q.
  - $P \Leftarrow Q$ If statement Q is true then so is statement P. P is a necessary condition for Q.
  - $P \Leftrightarrow Q$ Statement P is equivalent to statement Q. P is both a necessary and a sufficient condition for Q.
  - The **converse** of $P \Rightarrow Q$ is $P \Leftarrow Q$ or $Q \Rightarrow P$.
- **Proof by direct method**
  Step 1 Assume P to be true.
  Step 2 Demonstrate that Q is true as a consequence of the assumption in Step 1.
- **Proof by contradiction**
  Step 1 Assume that statement P is false.
  Step 2 Deduce from the assumption that there is a contradiction of some sort.
  Step 3 Infer that the assumption in Step 1 was wrong, which is equivalent to saying that P is true
- **Proof by counter-example**
  A general statement can be disproved by finding a single counter-example.

Formulae you must know

There are no specific formulae that you must know for this topic, but you do need to be aware of formulae from other topics as they may be used in questions on proof.

Don't make these mistakes ...

Don't forget that, in a proof, showing that a result is true for one particular value of a parameter does not necessarily mean that it is true for all parameters, so proving a special case does not prove the general case.

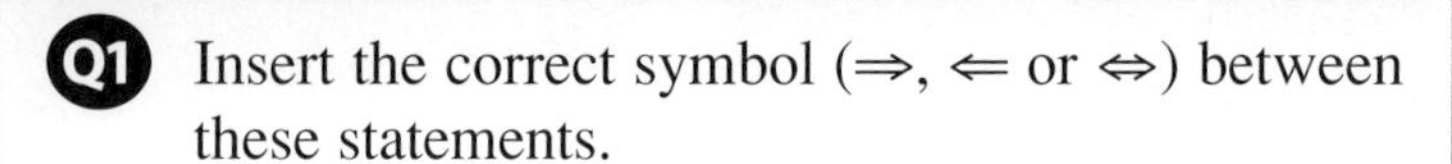

Exam Questions and Answers

How to score full marks

Q1 Insert the correct symbol ($\Rightarrow$, $\Leftarrow$ or $\Leftrightarrow$) between these statements.

(a) $\theta = \frac{\pi}{4}$ ☐ $\tan\theta = 1$

(b) $x^2 - 5x + 4 = 0$ ☐ $x = 4$

(c) $\mathbf{a.b} = 0$ ☐ $\mathbf{a}$ is perpendicular to $\mathbf{b}$

(d) $x^2 < 64$ ☐ $|x| < 8$

(a) $\theta = \frac{\pi}{4} \Rightarrow \tan\theta = 1$

- The converse statement $\tan\theta = 1 \Rightarrow \theta = \frac{\pi}{4}$ is only partially true because if $\tan\theta = 1$ then $\theta = \frac{\pi}{4} + n\pi$ for any integer n. So you can't write $\Leftrightarrow$.

(b) $x^2 - 5x + 4 = 0 \Leftarrow x = 4$

(c) a.b = 0 $\Leftarrow$ a is perpendicular to b

(d) $x^2 < 64 \Leftrightarrow |x| < 8$

- Again be careful because $x^2 - 5x + 4 = 0 \Rightarrow x = 4$ or $x = 1$.
- $\mathbf{a.b} = 0$ can also mean that $\mathbf{a} = 0$ or $\mathbf{b} = 0$ so you can't write $\Leftrightarrow$.

Q2 Prove the result:

'A triangle with sides that can be written in the form $n^2 + 1$, $n^2 - 1$, $2n$ (where $n > 1$) is right-angled.'

Show, by means of a counter example, that the converse is false.

For the triangle to be right-angled, it is necessary to show that:

$(n^2 - 1)^2 + (2n)^2 = (n^2 + 1)^2$

which is a statement of Pythagoras' theorem.

- For the first part, you can use a direct proof. Always state at the beginning the method you intend to use. Here you are using Pythagoras' theorem.

$$\begin{aligned} \text{LHS} &= (n^2 - 1)^2 + (2n)^2 \\ &= n^4 - 2n^2 + 1 + 4n^2 \\ &= n^4 + 2n^2 + 1 \\ &= (n^2 + 1)^2 \\ &= \text{RHS as required.} \end{aligned}$$

- For a direct proof you should give a clear algebraic argument to support your answer.

For a counter example, a triangle with sides of 5, 12 and 13 units is right-angled.

But $12 = 2n \Rightarrow n = 6$

Then $n^2 + 1 = 37$ and $n^2 - 1 = 35$

So this right-angled triangle does not have sides in the form $n^2 + 1$, $n^2 - 1$, $2n$.

- The converse statement says that for any right-angled triangle, the sides can be written as $n^2 + 1$, $n^2 - 1$, $2n$.
- For a counter example you just need to find one Pythagorean triple which is not of this form.

Q3 **(a)** Prove that a perfect square which is a multiple of 3 is the square of a multiple of 3.
(b) Prove that $\sqrt{3}$ is irrational.

(a) Any integer can be written in the form $3k$, $3k + 1$ or $3k + 2$.

So all perfect squares can be written in the form $(3k)^2$, $(3k + 1)^2$ or $(3k + 2)^2$.

$(3k + 1)^2 = 9k^2 + 6k + 1$
$= 3(3k^2 + 2k) + 1$
which is not a multiple of 3.

$(3k + 2)^2 = 9k^2 + 12k + 4$
$= 3(3k^2 + 4k + 1) + 1$
which is not a multiple of 3.

The only perfect square which is a multiple of 3 is $(3k)^2$ which is the square of a multiple of 3.

(b) Suppose that $\sqrt{3}$ is not irrational.

Then $\sqrt{3} = \frac{p}{q}$ where p and q are integers with no common factor.

$\sqrt{3} = \frac{p}{q} \Rightarrow p^2 = 3q^2$
$\Rightarrow p^2$ is a multiple of 3
$\Rightarrow p$ is a multiple of 3 (by (a))
$\Rightarrow p = 3k$
$\Rightarrow 9k^2 = 3q^2$
$\Rightarrow q^2 = 3k^2$
$\Rightarrow q^2$ is a multiple of 3
$\Rightarrow q$ is a multiple of 3
$\Rightarrow p$ and q have a common factor of 3

But this contradicts the supposition that and p and q have no common factor. Hence $\sqrt{3}$ is irrational.

How to score full marks

- Think carefully what you have to prove. Clearly the square of a multiple of 3 is a perfect square and it is also a multiple of 3, since $(3k)^2 = 9k^2 = 3(3k^2)$.

  You need to eliminate the cases that arise from squaring numbers which are not multiples of 3.

- Begin by stating the converse that $\sqrt{3}$ is rational and write it as a fraction.
- Use the method of proof by contradiction to prove that $\sqrt{3}$ is irrational.
- Now you need to show that this statement cannot hold.
- You need to argue the case carefully. Give a clear mathematical argument to support your answer.

Questions to try

Q1 Insert the correct symbol ($\Rightarrow$, $\Leftarrow$, $\Leftrightarrow$) between the following statements.

(a) $n = 2$	☐	$n^2 - n - 2 = 0$
(b) $ab = 0$	☐	$a = 0$ or $b = 0$
(c) n is a multiple of 2	☐	n is a multiple of 4
(d) $\sin\theta = \frac{1}{2}$	☐	$\theta = \frac{\pi}{6}$
(e) m is odd	☐	$m = 2n + 1$ for some integer n

Q2 State the converse of each of the following statements, and say whether the converse is true.

(a) If 4 divides n then 4 divides n^2, where n is a positive integer.
(b) For a polynomial $P(x)$, if $P(a) = 0$ then $(x - a)$ is a factor of $P(x)$.
(c) If $y = \sin x$ then $\frac{dy}{dx} = \cos x$.
(d) Every prime number greater than 3 is of the form $6n \pm 1$ where n is a positive integer.

Q3 Prove that if p is a prime number greater than 3 then $p^2 - 1$ is divisible by 24.

Q4 Prove that if n is an integer such that n^3 is even then n itself is even.

Q5 Prove that the cube root of 2 is not a rational number.

Q6 **(a)** Prove that every odd integer can be written in the form $4n + 1$ or $4n + 3$.
(b) Prove that the product of two integers of the form $4n + 1$ where $n > 0$ is itself of the form $4n + 1$.
(c) Find a counter example that shows the converse of the statement in (b) is not true.

Answers can be found on pages 92–93.

7 Coordinate geometry

Key points to remember

- The **equation of a circle**, centre (a, b) and of radius r, is $(x - a)^2 + (y - b)^2 = r^2$.
- The equation of a circle can also be expressed in the form $x^2 + y^2 + 2gx + 2fy + c = 0$.
- You can show that an equation in the form above is a circle by expressing it in the form $(x - a)^2 + (y - b)^2 = r^2$
  For example, $x^2 + y^2 - 4x - 8y + 16 = 0$ becomes $(x - 2)^2 + (y - 4)^2 = 2^2$, a circle of radius 2 with centre (2, 4).
- The **parametric equations** of a curve are functions $x = x(t)$ and $y = y(t)$, in which Cartesian coordinates of a point on the curve are given in terms of a third variable t called a **parameter**.
- For a circle, radius r, $x(t) = r\cos t$ and $y(t) = r\sin t$.
- The parameter t can often be eliminated from parametric equations. For example if $x = t^2$ and $y = \frac{1}{t^3}$, then $y = \frac{1}{x^{\frac{3}{2}}}$.

Formulae you must know

- $(x - a)^2 + (y - b)^2 = r^2$ is a circle centre (a, b) radius r.
- To complete the square,
  $x^2 + cx + d = (x + \frac{c}{2})^2 + (d - \frac{c^2}{4})$

Don't make these mistakes ...

Don't make simple algebraic mistakes when completing the square; watch out for the sign of $-\frac{c^2}{4}$ and remember to halve the coefficient of x.

Exam Questions and Answers

How to score full marks

Q1 **(a)** Find, in Cartesian form, an equation of the circle C with centre (1, 4) and radius 3.
(b) Determine by calculation, whether the point (2.9, 1.7) lies inside or outside C.

(a) $(x-1)^2 + (y-4)^2 = 3^2$
$x^2 - 2x + 1 + y^2 - 8y + 16 = 9$
$x^2 + y^2 - 2x - 8y + 8 = 0$

- Use the standard Cartesian formula for a circle, then expand and simplify.

(b) The distance from P to the centre is given by:
$d^2 = (2.9-1)^2 + (1.7-4)^2$
$= 8.9$
$d = 2.98$ and $2.98 < 3$
The point is inside the circle.

- If the point lies inside the circle, its distance from the centre of the circle must be smaller than the radius of the circle. Remember to give your answer in words.

Q2 A curve is defined in terms of a parameter t as and $x = 2\sec t$ and $y = 3\tan t$.
(a) Find the minimum distance of the curve from the origin.
(b) Show that $4y^2 - 9x^2 + 36 = 0$.

(a) $d = \sqrt{4\sec^2 t + 9\tan^2 t}$
$= \sqrt{4(1 + \tan^2 t) + 9\tan^2 t}$
$= \sqrt{4 + 13\tan^2 t}$
So the minimum value of d is 2.

- Use Pythagoras' theorem to calculate the distance of the point from the origin. The distance, d, is given by $d = \sqrt{x^2 + y^2}$. Then use the identity $1 + \tan^2 t = \sec^2 t$.

(b) $\sec t = \frac{x}{2}$

- Express $\sec t$ in terms of x.

$y = 3\tan t$
$y^2 = 9\tan^2 t$
$= 9(\sec^2 t - 1)$
$= 9((\frac{x}{2})^2 - 1)$
$= \frac{9x^2}{4} - 9$

- Express y in terms of $\sec t$, then substitute to eliminate t.

$4y^2 - 9x^2 + 36 = 0$

- Finally simplify to find the required result.

Q3 A circle, C with centre P, has equation $x^2 + y^2 - 10y + 16 = 0$.

(a) **(i)** Find the coordinates of P.

(ii) Show that the radius of the circle is 3.

(b) **(i)** The point $Q(\frac{12}{5}, \frac{16}{5})$ lies on the circle C.

The line L is the tangent to the circle at Q. Find the equation of L.

(ii) Verify that the point R(–3, –4) lies on the line L.

(c) Calculate the size of angle PRQ.

(a) $x^2 + y^2 - 10y + 16 = 0$

$(x - 0)^2 + (y - 5)^2 - 9 = 0$

$(x - 0)^2 + (y - 5)^2 = 9 = 3^2$

P has coordinates (0, 5) and the radius of the circle is 3.

(b) (i) The gradient of PQ is $\frac{5 - \frac{16}{5}}{0 - \frac{12}{5}} = -\frac{3}{4}$

The gradient of the tangent $= \frac{4}{3}$

The equation of the tangent at Q is $y - \frac{16}{5} = \frac{4}{3}(x - \frac{12}{5})$

$y = \frac{4}{3}x$

(ii) When $x = -3$, $y = \frac{4}{3} \times -3 = -4$

The point (–3, –4) lies on L.

(c)

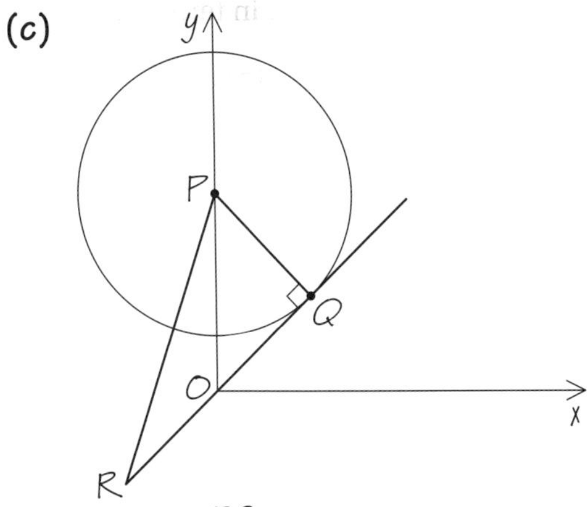

$\tan PRQ = \frac{PQ}{RQ}$

$RQ^2 = (-4 - \frac{16}{5})^2 + (-3 - \frac{12}{5})^2$

$RQ = 9$

$\tan PRQ = \frac{3}{9}$

angle PRQ = 18.4°

How to score full marks

- Rewrite the equation in standard form $(x - a)^2 + (y - b)^2 = r^2$ so that you can write down the centre (a, b) and radius r.
- PQ is a radius of the circle so PQ is perpendicular to the tangent. Find the gradient of PQ and then the gradient of the tangent is $-\frac{1}{\text{gradient of the tangent}}$.
- Substitute $x = -3$ into the equation of the tangent and show that at that point $y = -4$.
- A diagram shows that the triangle PRQ is right-angled. You can use simple trigonometry to find the value of angle PRQ.

Questions to try

Q1 Find the centre and radius of the circle with equation $x^2 + y^2 = 6x$.

The line $x + y = k$ is a tangent to this circle. Find the two possible values of the constant k, giving your answer in surd form.

Q2 Determine the coordinates of the centre C and the radius of the circle with equation $x^2 + y^2 + 4x - 10y + 13 = 0$.

Find the distance from the point P(2, 3) to the centre of the circle. Hence find the length of the tangents from P to the circle.

Q3 Find the range of values of m for which the line $y = mx$ and the circle $x^2 + y^2 - 6x - 8y + 24 = 0$:
(a) cut in two points
(b) touch at two points
(c) do not intersect.

Q4 The points(5, 5) and (–3, –1) are the ends of a diameter of the circle C with centre A. Write down the coordinates of A and show that the equation of C is $x^2 + y^2 - 2x - 4y - 20 = 0$.

The line L with equation $y = 3x - 16$ meets C at the points P and Q. Show that the x-coordinates of P and Q satisfy the equation $x^2 - 11x + 30 = 0$.

Hence find the coordinates of P and Q.

Q5 Write the following equation of a circle in standard form and hence find its centre and radius.

$x^2 + y^2 + 2x - 8y + 12 = 0$

Find the points of intersection of the circle and the line $y = 2x + 1$.

Q6 A penny-farthing bicycle on display in a museum is supported by a stand at points A and C. A and C lie on the front wheel.

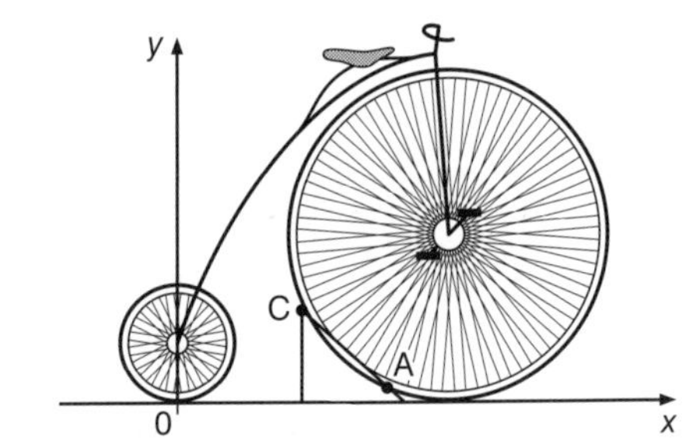

With coordinate axes as shown and 1 unit = 5 cm, the equation of the rear wheel (the small wheel) is $x^2 + y^2 - 6y = 0$ and the equation of the front wheel is $x^2 + y^2 - 28x - 20y + 196 = 0$.

(a) **(i)** Find the distance between the centres of the two wheels.
(ii) Hence calculate the clearance, i.e. the smallest gap, between the front and rear wheels. Give your answer to the nearest millimetre.

(b) B(7, 3) is halfway between A and C, and P is the centre of the front wheel.
(i) Find the gradient of PB.
(ii) Hence find the equation of AC and the coordinates of A and C.

Q7 A line is defined in terms of a parameter t by $x = 2t - 1$ and $y = 5t + 1$. Find the minimum distance of this line from the point with coordinates (8, 9).

Answers can be found on pages 94–95.

8 Differential equations

Key points to remember

- A **differential equation** is an equation involving **derivatives**, for example: $\frac{dy}{dx} = 4x - y$.
- A **first-order** differential equation only involves a first derivative. The general form of a first order differential equation is $\frac{dy}{dx} = f(x, y)$.
- If $f(x, y)$ is a function of x only then the differential equation can be solved by direct **integration**, for example if $\frac{dy}{dx} = 5x - 1$, then direct integration gives $y = \frac{5}{2}x^2 - x + c$.
- If $f(x, y)$ is in the form $u(x)v(y)$ then the variables are **separated**. The method of **separation of variables** can be used to give $\int \frac{1}{v(y)} dy = \int u(x)dx$. For example $\frac{dy}{dx} = xy$ becomes $\int \frac{1}{y} dy = \int xdx$.
- The **general solution** of a first-order differential equation contains one unknown constant and represents a family of solution curves, for example as shown on the right.

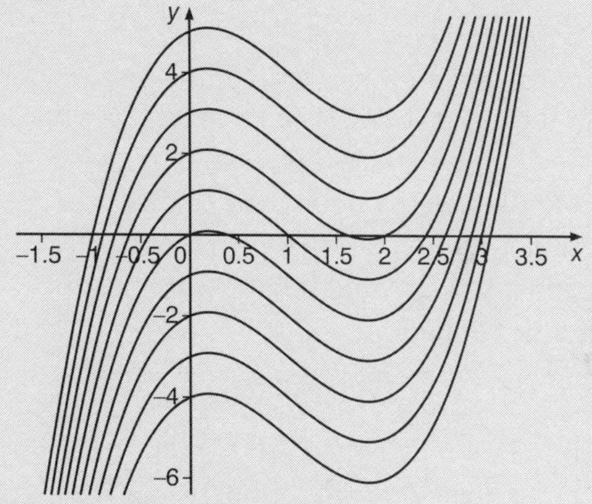

- A **particular solution** is one member of the family of solutions.
- Use a step-by-step method to find numerical solutions. If $\frac{dy}{dx} = f(x, y)$ and $y = y_0$ when $x = x_0$ then $y_{n+1} = y_n + hf(x_n, y_n)$ where h is the **step size**.

Formulae you must know

- You will need to use the laws of logarithms to simplify your answers.

  $\ln a + \ln b = \ln ab \qquad \ln a - \ln b = \ln\frac{a}{b}$

  If $\ln a = c$ then $a = e^c$
- You will need to know the integrals of basic functions.

Don't make these mistakes ...

Don't make a silly algebraic mistake in simplifying your answer. For example, if $\ln y = ax + c$ then $y = e^{ax + c} = e^{ax}e^c = Ae^{ax}$. A common mistake is to write $y = e^{ax} + e^c = e^{ax} + A$ which is wrong.

Don't make a mistake in separating the variables. For example, if $\frac{dy}{dx} = y + a$ you **must** divide by $(y + a)$ to give $\int \frac{1}{y+a}dy = \int 1dx$.

Q1 To control the pests inside a large greenhouse, 600 ladybirds were introduced. After t days there were P ladybirds in the greenhouse.

In a simple model, P is assumed to be a continuous variable satisfying the differential equation $\frac{dP}{dt} = kP$, where k is a constant and t is measured in days.

(a) Solve the differential equation, with initial condition $P = 600$ when $t = 0$, to express P in terms of k and t.

Observations of the number of ladybirds (estimated to the nearest hundred) were made as follows.

t	0	150	250
P	600	1200	3100

(b) Show that $P = 1200$ when $t = 150 \Rightarrow k \approx 0.004\,62$. Show that this is not consistent with the observed value when $t = 250$.

(c) In a refined model, allowing for seasonal variations, it is assumed that P satisfies the differential equation

$$\frac{dP}{dt} = P(0.005 - 0.008\cos 0.02t)$$

with initial condition $P = 600$ when $t = 0$.
Find an expression P in terms of t and comment on how well this fits with the data given above.
Show that, according to the refined model, the number of ladybirds will decrease initially and find the smallest number of ladybirds in the greenhouse.

How to score full marks

- Always read the question all the way through, before you start to answer it.

(a) $\frac{dP}{dt} = kP$

$\int \frac{1}{P} dP = \int k dt$

$\ln P = kt + c$

$\ln 600 = 0 + c$

$c = \ln 600$

$\ln P = kt + \ln 600$

$\ln \frac{P}{600} = kt$

$P = 600e^{kt}$

(b) $1200 = 600e^{150k}$

$e^{150k} = 2$

$k = \frac{1}{150}\ln 2 \approx 0.004\,62$

When $t = 250$ the model gives:

$P = 600e^{0.004\,62 \times 250}$

$= 1904$

The data gives $P = 3100$ so the model is not consistent with the data.

(c) $\frac{dP}{dt} = P(0.005 - 0.008\cos 0.02t)$

$\int \frac{1}{P} dP = \int (0.005 - 0.008\cos 0.02t) dt$

$\ln P = 0.005t - 0.4\sin 0.02t + c$

$\ln 600 = 0 - 0 + c$

$c = \ln 600$

$P = 600e^{0.005t - 0.4\sin 0.02t}$

When $t = 150$, $P = 1200$.

When $t = 250$, $P = 3073$.

The revised model fits the data well.

When $t = 0$, $\frac{dP}{dt} = -0.003P\ (< 0)$

so that initially the number of ladybirds is decreasing.

If $0.005 - 0.008\cos 0.02t = 0$

then $\cos 0.02t = \frac{5}{8}$

$t = 44.8$

When $t = 44.8$,

$P = 600e^{0.005 \times 44.8 - 0.4\sin 0.02 \times 44.8}$

$= 549$

The smallest number of ladybirds is 549.

How to score full marks 

- You are given the differential equation. Separate the variables by dividing each side by P. Integrate both sides and use the given initial condition to find the value of the constant c.
- Remember that $\ln P - \ln 600 = \ln \frac{P}{600}$ and if $\ln b = a$ then $b = e^a$.
- Substitute $P = 1200$ and $t = 150$ to find a value of k.
- Show that the model and data do not agree when $t = 250$. State that the model is not consistent – do not leave the examiner to interpret your solution.
- Use the given initial conditions to find the value of c.
- Find values of P for $t = 150$ and 250, then interpret your solution.
- To find the initial change in population find $\frac{dP}{dt}$ when $t = 0$.
- The negative value implies that the number of ladybirds is decreasing. The minimum number of ladybirds occurs when $\frac{dP}{dt} = 0$. Find when this occurs and substitute this value of t into the equation, to find the smallest value of P.

Q2 Solve the differential equation:

$$\frac{dy}{dx} = \sqrt{y}\sec^2 3x$$

given that $y = 1$ when $x = 0$, expressing your answer in the form $y = f(x)$.

$$\frac{dy}{dx} = \sqrt{y}\sec^2 3x$$

$$\int \frac{1}{\sqrt{y}}\,dy = \int \sec^2 3x\,dx$$

$$2\sqrt{y} = \tfrac{1}{3}\tan 3x + c$$

When $x = 0$, $y = 1$.

$$2 = 0 + c$$

$$2\sqrt{y} = \tfrac{1}{3}\tan 3x + 2$$

$$\Rightarrow y = (1 + \tfrac{1}{6}\tan 3x)^2$$

How to score full marks

- Divide both sides by $\sqrt{y}$ and integrate each side.
- Use the given condition to find the value of the constant c.

Questions to try

Q1 In a chemical reaction, hydrogen peroxide is converted into water and oxygen. At time t after the start of the reaction, the quantity of hydrogen peroxide that has **not** been converted is x and the rate at which x is decreasing is proportional to x.

(a) Write down a differential equation involving x and t.

(b) Given that $x = x_0$ initially, show that:

$$\ln\frac{x}{x_0} = -kt$$

where k is a positive constant.

(c) In an experiment, the time taken for the hydrogen peroxide to be reduced to half of its original quantity was 3 minutes. Find, to the nearest minute, the time that would be required to reduce the hydrogen peroxide to one-tenth of its original quantity.

(d) **(i)** Express x in terms of x_0 and t.

(ii) Sketch a graph showing how x varies with t.

Q2 A cylindrical tank with a horizontal circular base is leaking. At time t minutes the depth of oil in the tank is h metres. It is known that $h = 10$ when $t = 0$ and that $h = 5$ when $t = 40$.

Alan assumes that the rate of change of h with respect to t is constant.

(a) Find an expression for h in terms of t.

Bhavana assumes that the rate of change of h with respect to t is proportional to h.

(b) Form a differential equation and, using the conditions given, solve it to find Bhavana's expression for h in terms of t.

(c) Find, in each case, the value of h when $t = 60$.

(d) Briefly explain which assumption you would use.

Questions to try

Q3 A cylindrical container has a height of 200 cm. The container was initially full of a chemical but there is a leak from a hole in the base. When the leak is noticed, the container is half-full and the level of the chemical is dropping at a rate of 1 cm per minute. It is required to find the time (in minutes) for which the container has been leaking. To model the situation it is assumed that, when the depth of the chemical remaining is x cm, the rate at which the level is dropping is proportional to $\sqrt{x}$. Set up and solve an appropriate differential equation, and hence show that the container has been leaking for about 80 minutes.

Q4 Solve the differential equation:

$$x^2\frac{dy}{dx} = y + 1$$

given that $y = 0$ when $x = 1$. Express your answer in the form $y = f(x)$.

Q5 Solve the differential equation:

$$\frac{dv}{dt} = 10 - 3v$$

where v is the speed of a particle at time t, given that $v = 2$ when $t = 0$. Find the limiting value of v.

Q6 The rate at which a rumour spreads in a crowd of people can be modelled with the differential equation:

$$\frac{dn}{dt} = 0.0002n(N - n)$$

where N is the number in the crowd and n is the number who have heard the rumour at time t minutes. There are 5000 people in the crowd and initially 10 of them have heard the rumour.

(a) Copy and complete the table below for a step-by-step solution of the differential equation and estimate the number of people who have heard the rumour after 10 minutes.

n	t	$\frac{dn}{dt}$	dt	dn
10	0		5	
	5			
	10			

(b) State a means by which your estimate could be improved.

Answers can be found on pages 96–97.

9 Partial fractions and integration

Key points to remember

- **Partial fractions** are used to express algebraic fractions as the sum of a number of simpler fractions, for example,

  $$\frac{3}{(x+6)(x-9)} = \frac{A}{x+6} + \frac{B}{x-9}$$

  $$\frac{5x}{(x+1)^2(x-5)} = \frac{A}{x+1} + \frac{B}{(x+1)^2} + \frac{C}{x-5}$$

  or if the highest power in the denominator is the same as the numerator:

  $$\frac{x^2}{(x-1)(x+3)} = A + \frac{B}{x-1} + \frac{C}{x+3}$$

- Solve partial fraction problems by forming a single fraction, for example:

  $$\frac{3}{(x+6)(x-9)} = \frac{A}{x+6} + \frac{B}{x-9} = \frac{A(x-9)+B(x+6)}{(x+6)(x-9)} \Rightarrow A(x-9)+B(x+6) = 3$$

  Substitute $x = -6$ to find A.

  Substitute $x = 9$ to find B.

- Alternatively, solve as simultaneous equations.
- Partial fractions are usually easy to integrate. Use the laws of logarithms to simplify answers.

  $$\int_1^2 \left(\frac{2}{x+1} + \frac{3}{x+2}\right)dx = \Big[2\ln(x+1) + 3\ln(x+2)\Big]_1^2 = 4\ln 2 - \ln 3$$

Formulae you must know

- For every distinct linear factor $(ax+b)$ in the denominator you should include a term $\frac{A}{ax+b}$.
- For every repeated factor in the denominator you should include the two terms $\frac{A}{ax+b} + \frac{B}{(ax+b)^2}$.
- The laws of logarithms are useful for simplifying answers.

  $\ln a + \ln b = \ln ab \quad \ln a - \ln b = \ln \frac{a}{b} \quad n\ln a = \ln a^n$

Don't make these mistakes ...

Don't omit a term when deciding on the form of the partial fractions.

Don't make errors when integrating.
Note $\int \frac{A}{ax+b}dx = \frac{A}{a}\ln(ax+b) + c$
and $\int \frac{B}{(ax+b)^2}dx = -\frac{B}{a(ax+b)}$.

Don't make a silly algebraic or arithmetic error when forming or solving simultaneous equations or when substituting.

Exam Questions and Answers

How to score full marks

Q1 Find $\int_0^2 \frac{x}{(x+1)(x+2)}\,dx$, giving your answer as a single logarithm.

$$\frac{x}{(x+1)(x+2)} = \frac{A}{x+1} + \frac{B}{x+2}$$

$$= \frac{A(x+2) + B(x+1)}{(x+1)(x+2)}$$

$$= \frac{(A+B)x + (2A+B)}{(x+1)(x+2)}$$

- First, decide on the form of the partial fractions. As there are no repeated factors in the denominator, this is relatively straightforward.

$A + B = 1$

$2A + B = 0$

$A = -1$

$B = 1 + 1$

$= 2$

- Convert from the partial fraction form to a single fraction and compare the coefficients with the original fraction. Form a pair of simultaneous equations and solve them to find the values of A and B.

$$\int_0^2 \frac{x}{(x+1)(x+2)}\,dx$$

$$= \int_0^2 \left(\frac{-1}{x+1} + \frac{2}{x+2}\right)dx$$

$$= \Big[-\ln(x+1) + 2\ln(x+2)\Big]_0^2$$

- Now integrate.

$$= (-\ln 3 + 2\ln 4) - (-\ln 1 + 2\ln 2)$$

$$= \ln\frac{16}{3 \times 4}$$

$$= \ln\frac{4}{3}$$

- Use the laws of logarithms to convert the answer to a single logarithm.

Alternative method for partial fractions

$$\frac{x}{(x+1)(x+2)} = \frac{A(x+2) + B(x+1)}{(x+1)(x+2)}$$

$x = A(x+2) + B(x+1)$

If $x = -1$: $\quad -1 = A$

$\Rightarrow A = -1$

If $x = -2$: $\quad -2 = -B$

$\Rightarrow B = 2$

- Alternatively, use the substitution method. Because there is a term in $(x + 1)$, substitute $x = -1$ to find A, then because there is a term in $(x + 2)$ substitute $x = -2$ to find B.

Q2 Find $\int_2^3 \frac{2x+1}{x(x+1)^2}dx$, giving your answer in the form $a + \ln b$.

$$\frac{2x+1}{x(x+1)^2}$$

$$= \frac{A}{x} + \frac{B}{x+1} + \frac{C}{(x+1)^2}$$

$$= \frac{A(x+1)^2 + Bx(x+1) + Cx}{x(x+1)^2} \quad (1)$$

$$= \frac{(A+B)x^2 + (2A+B+C)x + A}{(x+1)(x+2)} \quad (2)$$

Compare the numerators.

If $x = 0$: $1 = A \Rightarrow A = 1$

If $x = -1$: $-1 = -C \Rightarrow C = 1$

Comparing the coefficients of x^2:

$A + B = 0$

$\Rightarrow B = -1$

$$\int_2^3 \frac{2x+1}{x(x+1)^2}dx$$

$$= \int_2^3 \left(\frac{1}{x} - \frac{1}{x+1} + \frac{1}{(x+1)^2}\right)dx$$

$$= \left[\ln x - \ln(x+1) - \frac{1}{x+1}\right]_2^3$$

$$= (\ln 3 - \ln 4 - \frac{1}{4}) - (\ln 2 - \ln 3 - \frac{1}{3})$$

$$= \frac{1}{12} + \ln(\frac{3 \times 3}{4 \times 2})$$

$$= \frac{1}{12} + \ln\frac{9}{8}$$

How to score full marks 

- Because the denominator has a repeated factor, you need three terms. Combine them into a single fraction.
- Substitute $x = 0$ to find A.
- Substitute $x = -1$ to find C.
- To find B, form an equation by comparing the coefficients of x^2.
- Now integrate. Note that the third term does not give a ln when integrated.
- Use the laws of logarithms to obtain the answer in the specified format.

Questions to try

Q1 Find $\int_0^1 \frac{x}{(x+2)(x+3)}\,dx$.

Q2 Express $\frac{15-13x+4x^2}{(1-x)^2(4-x)}$ in partial fractions.

Hence show that $\int_2^3 \left(\frac{15-13x+4x^2}{(1-x)^2(4-x)}\right)dx = 1 + \ln 4$.

Q3 $f(x) = \frac{x+4}{(x+1)^2(x+2)}$

(a) Express f(x) in the form:

$$\frac{A}{(x+1)^2} + \frac{B}{x+1} + \frac{C}{x+2}$$

where the constants A, B and C are to be found.

(b) Evaluate f'(1), giving your answer as an exact rational number.

The finite region R is bounded by the curve with equation $y = f(x)$, the coordinate axes and the line $x = 3$.

(c) Find the area of R, giving your answer in the form $p + \ln q$, where p and q are rational numbers to be found.

Q4 Express $\frac{2-x-2x^2}{(1-x)(1-2x)^2}$ as the sum of three partial fractions.

Q5 **(a)** Express $\frac{18}{x^2(x+3)}$ in the form $\frac{A}{x} + \frac{B}{x^2} + \frac{C}{(x+3)}$ and state the values of the constants A, B and C.

(b) Hence show that $\int_1^3 \frac{18}{x^2(x+3)}\,dx = 4 - 2\ln 2$.

Q6 Find $\int_2^4 \frac{4x+3}{x(x+1)^2}\,dx$.

Answers can be found on pages 98–99.

10 Moments and centres of mass

Key points to remember

- To find the **centre of mass** of a composite body treat it as two or more particles and use the formula:

  $$\bar{x} = \frac{\Sigma m_i x_i}{\Sigma m_i}$$

  and the corresponding formula for $\bar{y}$.
- When a body is suspended in **equilibrium** its centre of mass is below the point of suspension.

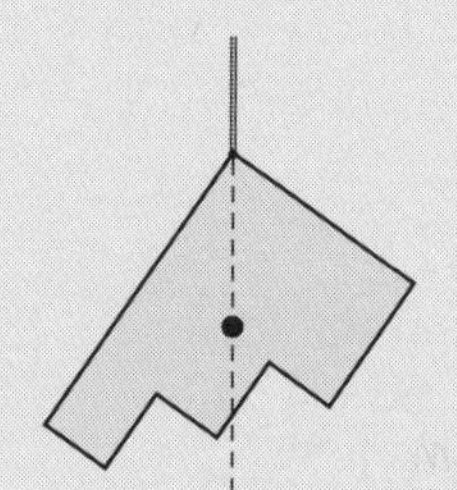

- When a body on a slope is on the point of **toppling**, its centre of mass is directly above the point about which it would rotate.

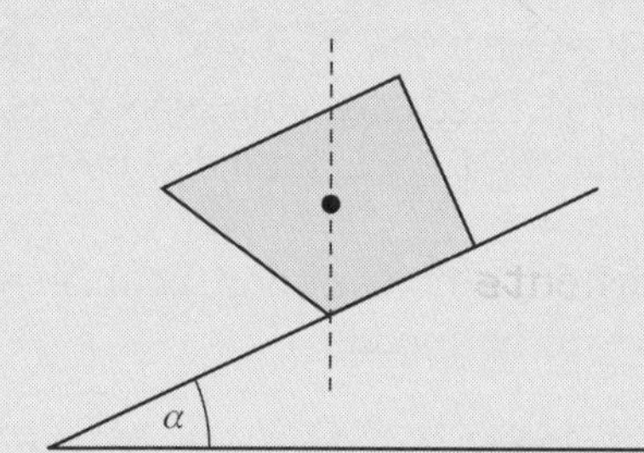

- The **moment** of a force about a point is the product of the magnitude of the force, F, and the perpendicular distance, d, of the point from the line of action of the force.

  Moment of F about O = $Fd = Fa\sin\alpha$

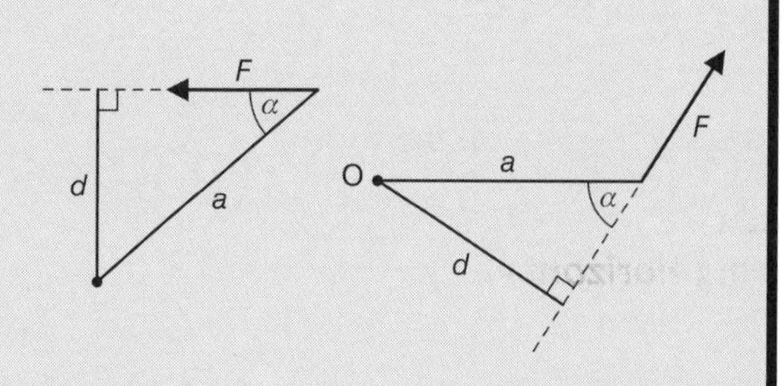

- In equilibrium:
  - the resultant force is zero **and**
  - the total moment is zero.

Formulae you must know

- $\bar{x} = \frac{\Sigma m_i x_i}{\Sigma m_i}$
- $\bar{y} = \frac{\Sigma m_i y_i}{\Sigma m_i}$
- Moment = force × perpendicular distance

Don't make these mistakes ...

- Don't make silly arithmetic errors when calculating centres of mass.
- Don't use the wrong distances when calculating moments or positions of centres of mass.
- Don't try to avoid using a diagram, when you really need one.

 A uniform rod of mass 16 kg and length 3 metres leans against a smooth vertical wall. The other end of the rod is on rough horizontal ground. The coefficient of friction between the ground and the rod is 0.5. If the rod is on the point of slipping, find the angle between the rod and the ground.

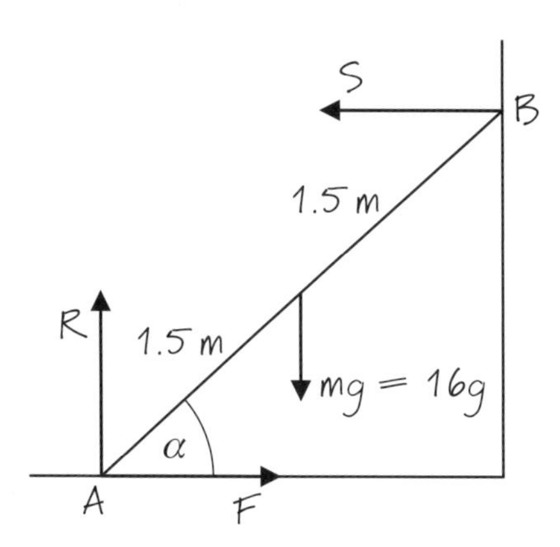

- Drawing a diagram helps to identify the forces acting and the distances to use when taking moments. Take moments about the bottom of the rod to find S.

Taking moments about the bottom of the ladder:

$1.5\cos\alpha \times 16 \times 9.8 = 3\sin\alpha \times S$

$S = \dfrac{1.5 \times 16 \times 9.8\cos\alpha}{3\sin\alpha}$

$= \dfrac{78.4}{\tan\alpha}$

Resolving horizontally:

$F = S$

$= \dfrac{78.4}{\tan\alpha}$

- Note that $F = S$, as the horizontal forces are in equilibrium.

Resolving vertically:

$R = mg$

$= 16g$

$= 16 \times 9.8$

$= 156.8$

- $R = 16g$ as the vertical forces are in equilibrium.

$F = \mu R$

$\dfrac{78.4}{\tan\alpha} = 0.5 \times 156.8$

$\tan\alpha = 1$

$\alpha = 45°$

- Then use the friction equation, with $F = \mu R$, as the rod is on the point of sliding, and find the value of α.

How to score full marks

Q2 A bridge is made from two planks joined together as shown in the diagram. The bottom plank has length 2 m and mass 20 kg. The top plank has length 1.4 m and mass 10 kg.

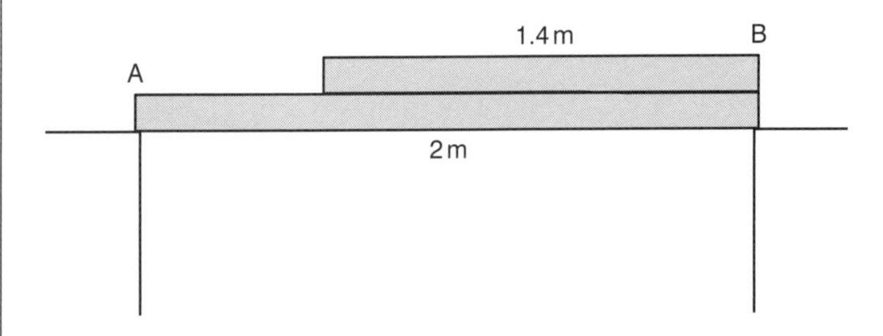

(a) Find the distance of the centre of mass from the point A.

(b) Calculate the magnitudes of the reaction forces that act at A and B.

- To calculate the position of the centre of mass, you need to note that the centre of the top plank is 1.3 m from A.

(a) $\bar{x} = \dfrac{20 \times 1 + 10 \times 1.3}{20 + 10} = 1.1\text{ m}$

(b) $2 \times R_B = 30g \times \bar{x} = 30 \times 9.8 \times 1.1$

$R_B = 161.7\text{ N}$

- Take moments about A to find the reaction forces at B. The clockwise moment due to the weight will be balanced by an anti-clockwise moment due to the reaction force.

$R_A + 161.7 = 30g = 30 \times 9.8$

$R_A = 132.3\text{ N}$

- The sum of the reaction forces is equal to the weight, so you can now calculate the reaction at A.

Q3 The diagram shows a uniform lamina, made up of two rectangles. It is suspended from the corner B and hangs in equilibrium. Find the angle between AB and the vertical.

- You will need to use the standard formulae, but with areas instead of masses. You can do this, as the mass is proportional to the area. You should treat the body as two rectangles.

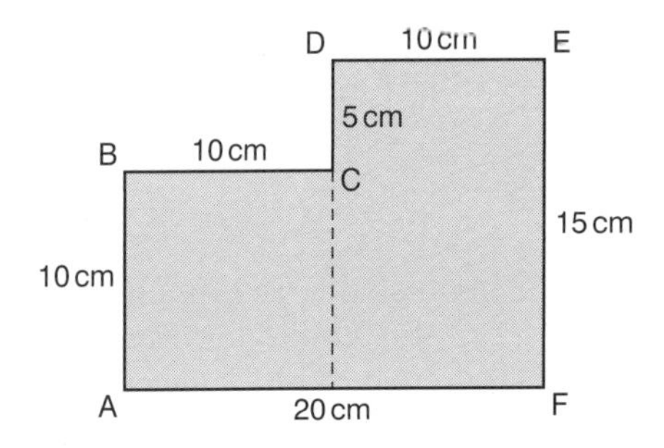

$$\bar{x} = \frac{100 \times 5 + 150 \times 15}{250} = 11\text{ m}$$

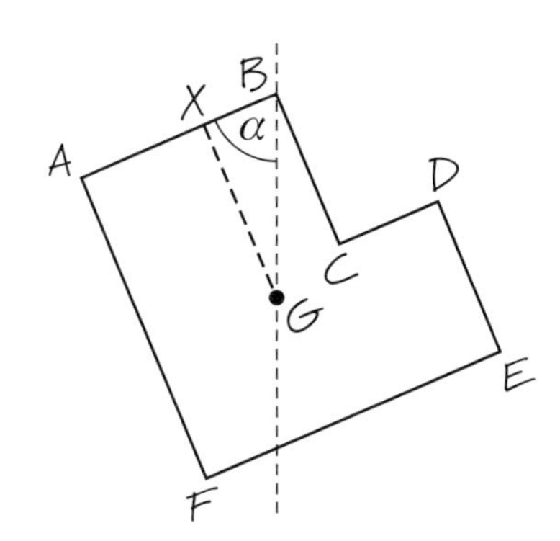

$$\bar{y} = \frac{100 \times 5 + 150 \times 7.5}{250} = 6.5\text{ m}$$

$$\tan\alpha = \frac{GX}{BX} = \frac{11}{10 - 6.5} = \frac{11}{3.5}$$

$$\alpha = 72.3°$$

How to score full marks

- Note that $\bar{x}$ gives the distance of the centre of mass from AB and $\bar{y}$ gives the distance from AF.
- A clear diagram is important to find the angle.
- Note that in this case you need to calculate the length *BX*, not just $\bar{x}$ and $\bar{y}$.

Questions to try

Q1 A lollipop stick of a children's crossing patrol warden can be modelled by a uniform rod AB together with a uniform circular disc attached to the rod at B. The diameter BC of the disc is in the same straight line as AB, as shown in the diagram. The rod is of length 1.4 m and weight 20 N and the disc is of diameter 0.4 m and weight 10 N.

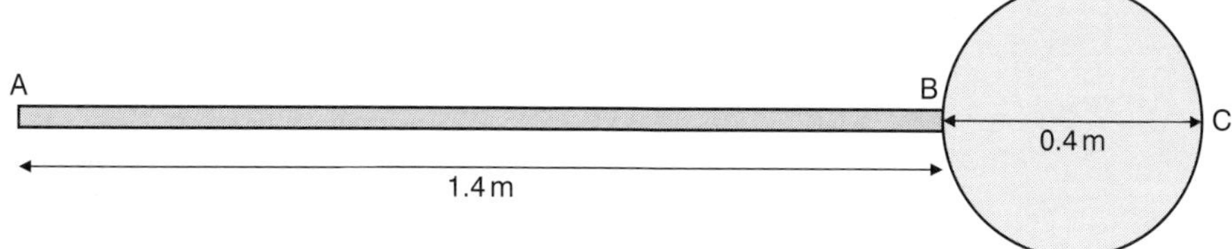

Show that the centre of mass of the lollipop stick is 1 m from A.

Q2 A uniform ladder, of mass m and length $2a$, has one end on rough horizontal ground. The other rests against a smooth vertical wall. A man of mass $3m$ stands at the top of the ladder and the ladder is in equilibrium. The coefficient of friction between the ground and the ladder is $\frac{1}{4}$, and the ladder makes an angle α with the wall, as shown in the diagram. The ladder is in a vertical plane perpendicular to the wall.

Show that $\tan\alpha \leqslant \frac{2}{7}$.

Questions to try

A uniform plane lamina ABCDE is formed by joining a uniform square ABDE with a triangular lamina BCD, of the same material, along the side BD, as shown in the diagram. The lengths AB, BC and CD are 18 cm, 15 cm and 15 cm respectively.

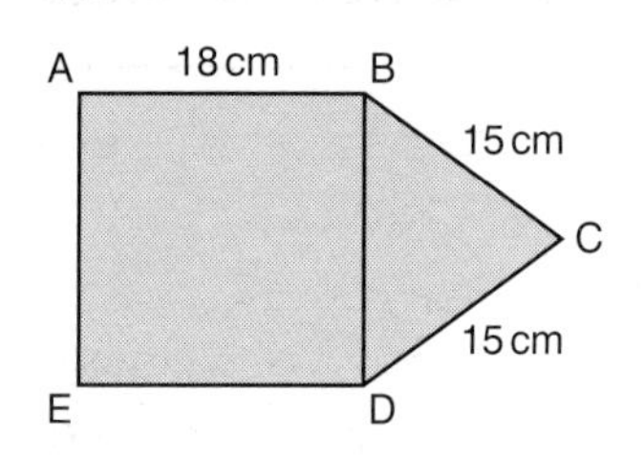

(a) Find the distance of the centre of mass of the lamina from AE.

The lamina is freely suspended from B and hangs in equilibrium.

(b) Find, in degrees to one decimal place, the angle that BD makes with the vertical.

The diagram shows a uniform lamina.

The lamina is suspended from the corner A. Find the angle between the side AB and the vertical when the lamina is at rest.

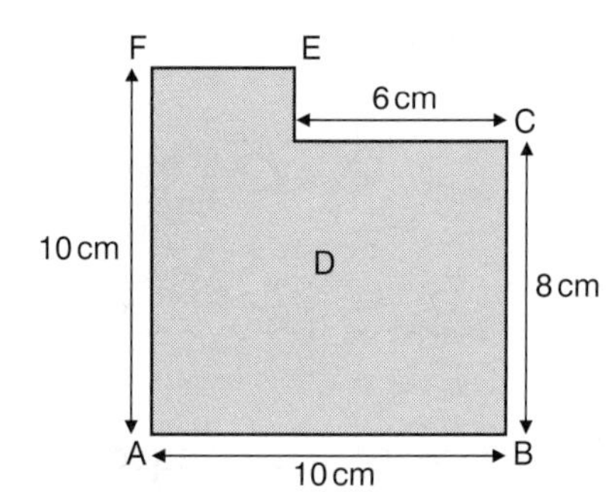

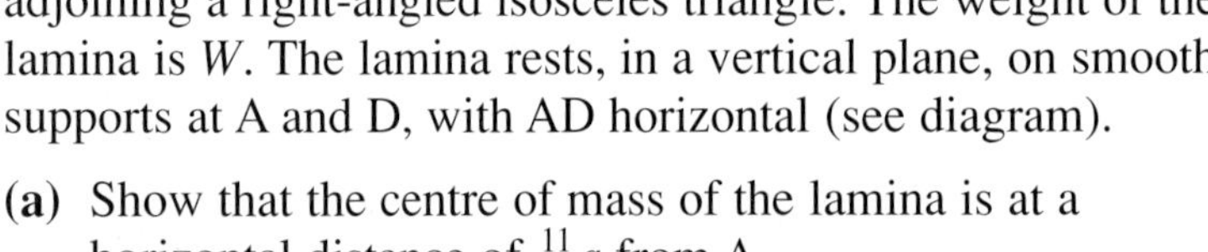

A uniform lamina ABCD has the shape of a square of side a adjoining a right-angled isosceles triangle. The weight of the lamina is W. The lamina rests, in a vertical plane, on smooth supports at A and D, with AD horizontal (see diagram).

(a) Show that the centre of mass of the lamina is at a horizontal distance of $\frac{11}{9}a$ from A.

(b) Find, in term of W, the magnitudes of the forces on the supports at A and D.

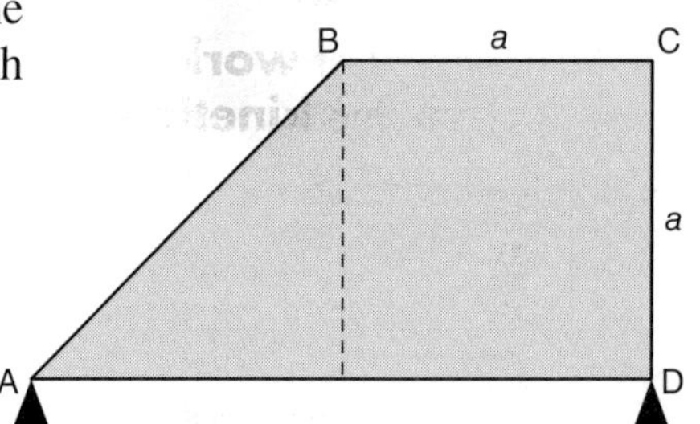

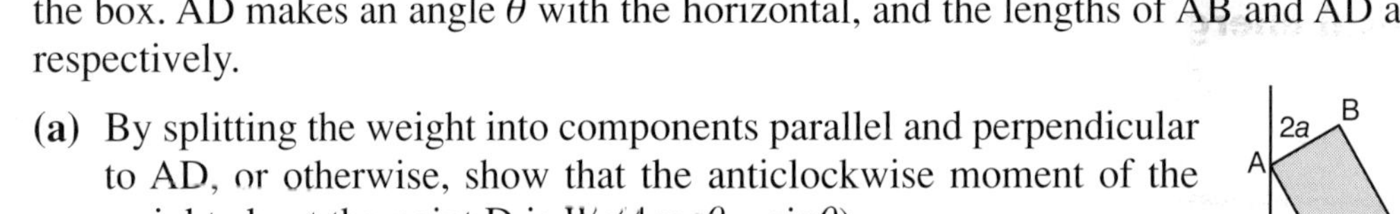

A uniform rectangular box of weight W stands on a horizontal floor and leans against a vertical wall. The diagram shows the vertical cross-section ABCD containing the centre of mass G of the box. AD makes an angle θ with the horizontal, and the lengths of AB and AD are $2a$ and $8a$ respectively.

(a) By splitting the weight into components parallel and perpendicular to AD, or otherwise, show that the anticlockwise moment of the weight about the point D is $Wa(4\cos\theta - \sin\theta)$.

(b) The contact at A between the box and the wall is smooth. Find, in terms of W and θ, the magnitude of the force acting on the box at A.

(c) The contact at D between the box and the ground is rough, with coefficient of friction μ. Given that the box is about to slip, show that $\tan\theta = \dfrac{4}{8\mu + 1}$.

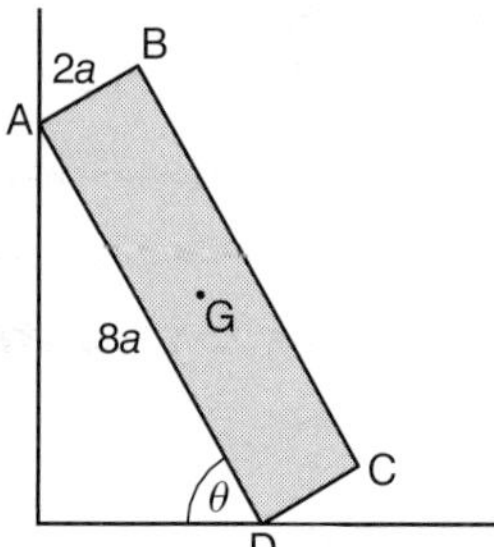

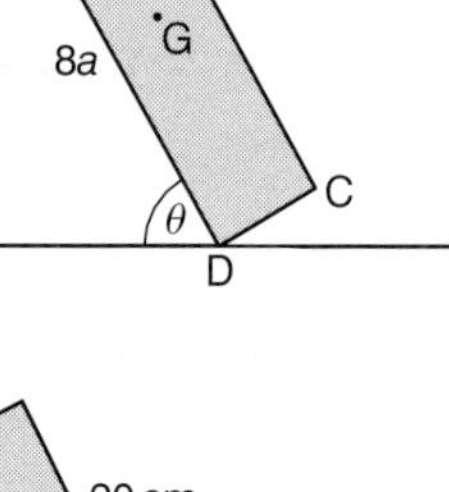

The diagram shows the cross-section of a uniform solid block. This cross-section has dimensions 20 cm by 10 cm and lies in a vertical plane. The block rests in equilibrium on a rough plane of which the inclination α to the horizontal can be varied. The coefficient of friction between the block and the plane is 0.7. Given that α is gradually increased from zero, determine whether equilibrium is broken by sliding or toppling.

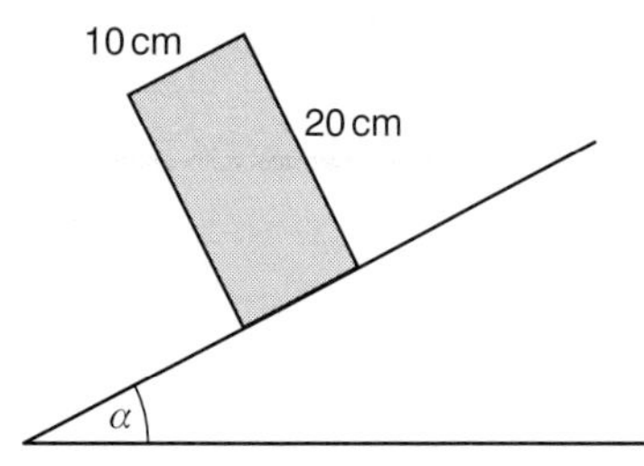

Answers can be found on pages 100–101.

11 Energy

Key points to remember

- Every moving body has **kinetic energy**.
  Calculate kinetic energy as KE = $\frac{1}{2}mv^2$.
- **Work** is done by a force F when it moves an object a distance d.
  Calculate work done as Fd when the force acts in the direction of motion, or $Fd\cos\theta$ when the force acts at an angle θ to the direction of motion.

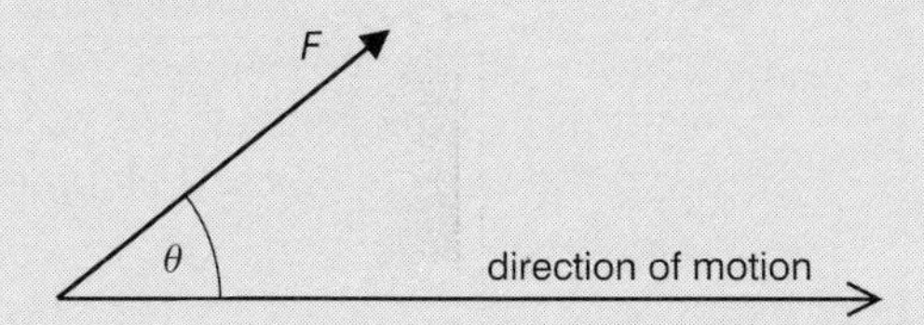

- The work done is equal to the **change in kinetic energy**.
- If the amount of **work** done is **negative**, the **kinetic energy** of the body is **reduced**.
- The **gravitational potential energy** depends on the **mass** of the object and its **height** from a fixed (the ground).
  Calculate gravitational potential energy as PGE = mgh.
- When no resistance or friction forces act, **mechanical energy is conserved**.
- **Power** is the rate of doing work.
  Calculate power as $P = Fv$.

Don't make these mistakes …

Don't use the wrong angle when working with forces that are not acting in the direction of motion.

Don't forget to include negative signs when calculating the total work done, when there are forces that oppose the motion of body under consideration.

Don't forget to include resistance forces.

Don't try to use $P = Fv$ without first finding the driving force that is acting.

Formulae you must know

- KE = $\frac{1}{2}mv^2$
- Work = Fd or $Fd\cos\theta$
- GPE = mgh
- $P = Fv$

Exam Questions and Answers

How to score full marks

Q1 A bullet, of mass 60 grams, travelling horizontally at $120\,\text{m s}^{-1}$ hits a wooden post. As it passes through the post its speed is reduced to $50\,\text{m s}^{-1}$. The thickness of the post is 15 cm.

(a) Find the magnitude of the average force that the post exerts on the bullet.

(b) An identical bullet is then fired into a thicker post and comes to rest inside the post. How far does the bullet penetrate the post?

(a) KE lost

$= \frac{1}{2} \times 0.06 \times 120^2 - \frac{1}{2} \times 0.06 \times 50^2$

$= 357\text{ J}$

$0.15R = 357$

$R = 2380\text{ N}$

(b) $\frac{1}{2} \times 0.06 \times 120^2 = 2380d$

$d = \frac{432}{2380} = 0.182\text{ m}$

- As the bullet passes through the wood a resistance force will act on the bullet. This resistance force will reduce the kinetic energy of the bullet.
- First calculate the kinetic energy lost. This will be equal to the work done against the resistance force. Since you know that the bullet travels 15 cm through the wood, you can calculate the average force, R.
- In the second case all the kinetic energy will be lost as work is done against the resistance force.
- Assuming that the force has magnitude 2380 N, you can calculate the distance that the bullet penetrates the wood.

Q2 A trolley has mass 50 kg. It rolls down a slope inclined at 10° to the horizontal. It is initially moving at $2\,\text{m s}^{-1}$ and reaches a speed of $v\,\text{m s}^{-1}$ after travelling 100 m.

(a) Find v if no resistance forces act on the trolley.

(b) Given that $v = 12$ and that a constant resistance force acts on the trolley, find the magnitude of the resistance force.

(a) Work done by gravity

$= 50 \times 9.8 \times 100 \times \cos 80° = 8509\text{ J}$

$8509 = \frac{1}{2} \times 50v^2 - \frac{1}{2} \times 50 \times 2^2$

$v^2 = \frac{8509 + 100}{25} \Rightarrow v = 18.6\text{ m s}^{-1}$

(b) Gain in KE

$= \frac{1}{2} \times 50 \times 12^2 - \frac{1}{2} \times 50 \times 2^2 = 3500\text{ J}$

Work done against resistance

$= 8509 - 3500 = 5009\text{ J}$

Resistance force $= \frac{5009}{100}$

$= 50.1\text{ N (to 3 s.f.)}$

- First calculate the work done by gravity, as this is the only force that does work in this case. Then use work done = change in KE, to find v.
- In the second case, find the gain in kinetic energy first. Then subtract this from the work done by gravity to find the work done against the resistance force. You can divide this by the distance travelled, to give the magnitude of the force.

Q3 A car moving at $v\,\text{m s}^{-1}$ experiences a resistance force of magnitude $40v\,\text{N}$. The car has mass 1200 kg and a maximum power output of 40 000 W.

(a) Find the maximum possible acceleration of the car while it travels at $16\,\text{m s}^{-1}$ horizontally.

(b) Find the maximum speed of the car down a slope inclined at an angle α to the horizontal where $\sin\alpha = \frac{1}{20}$.

(a) $F - 40 \times 16 = 1200a \Rightarrow F = 1200a + 640$

$P = Fv$

$40\,000 = (1200a + 640) \times 16$

$2500 = 1200a + 640$

$a = 1.55\,\text{m s}^{-2}$

(b) $F + 1200 \times 9.8 \times \sin\alpha = 40v$

$F = 40v - 1200 \times 9.8 \times \frac{1}{20}$

$= 40v - 588$

$P = Fv$

$40\,000 = (40v - 588)v$

$0 = 40v^2 - 588v - 40\,000$

$$v = \frac{588 \pm \sqrt{588^2 - 4 \times 40 \times -40\,000}}{2 \times 40}$$

$v = 39.8$ or -25.1 so $v = 39.8\,\text{m s}^{-1}$

How to score full marks

- Both parts of this question depend on finding the forward force acting on the car and then applying $P = Fv$.
- Find the resultant force on the car and apply Newton's second law. Once F is the subject of the expression apply $P = Fv$ and solve the equation for a.
- The driving force, F, and the component of gravity down the slope are balanced by the resistance force when the car moves at its maximum speed. This enables you to find F.
- Use $P = Fv$ to find a quadratic equation and solve it using the standard formula.
- You need the positive solution.

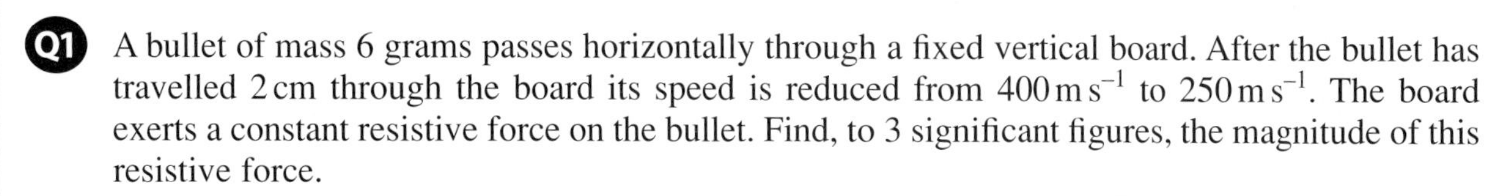

Questions to try

Q1 A bullet of mass 6 grams passes horizontally through a fixed vertical board. After the bullet has travelled 2 cm through the board its speed is reduced from $400\,\text{m s}^{-1}$ to $250\,\text{m s}^{-1}$. The board exerts a constant resistive force on the bullet. Find, to 3 significant figures, the magnitude of this resistive force.

Q2 A soldier, of mass 80 kg, swings on a rope of length 6 m. He is to be modelled as a particle that describes a circular arc from A, through B to C. The path is shown in the following diagram. The point A is 2 m higher than B. Initially the soldier moves at $2\,\text{m s}^{-1}$ at A and in a direction perpendicular to the rope.

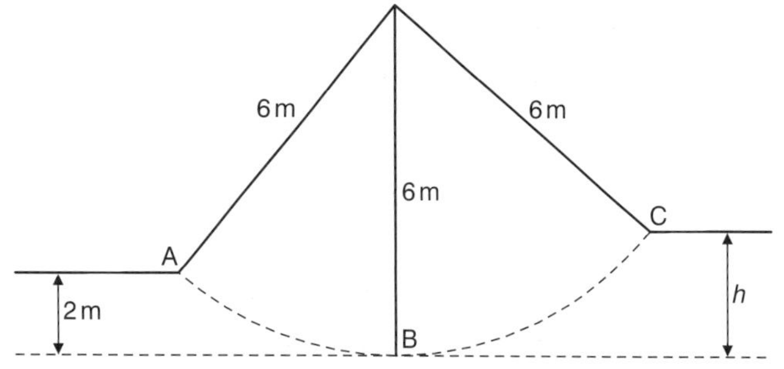

(a) Find the kinetic energy and velocity of the soldier at B, stating any assumptions that you make.

(b) Find h, if the soldier comes to rest at C before swinging back.

(c) Explain why the tension does no work in this situation.

Questions to try

Q3 As a van, of mass 1200 kg, skids 25 m, on a horizontal surface, its speed is halved from $30\,\text{m s}^{-1}$ to $15\,\text{m s}^{-1}$.

(a) Find the kinetic energy lost by the van as it skids.
(b) If the coefficient of friction between the van's tyres and the road is 0.8, find the work done against the friction force.
(c) Assuming that the air resistance force is constant, find out how much further the van travels before it stops.

Q4 A ball, of mass 200 grams, is dropped from a height of h m. The ball is initially at rest. It is travelling at a speed of $3\,\text{m s}^{-1}$ when it hits the ground. As it falls a constant resistance force of magnitude 0.8 N acts on the ball.

(a) Find h.
(b) Given that the ball rebounds to a height 0.3 m, find the speed at which it leaves the ground.

Q5 A straight road is inclined at an angle α to the horizontal, where $\sin\alpha = \frac{1}{20}$. A lorry of mass 4800 kg moves up the road at a constant speed of $12\,\text{m s}^{-1}$. The non-gravitational resistance to the motion of the lorry is constant and has magnitude 2000 N.

(a) Find, in kW, to 3 significant figures, the rate of working of the lorry's engine.

The road becomes horizontal. The lorry's engine continues to work at the same rate and the resistance to motion remains the same. Find:

(b) the acceleration of the lorry immediately after the road becomes horizontal
(c) the maximum speed, in m s^{-1} to 3 significant figures, at which the lorry will go along the horizontal road.

Q6 A car of mass 650 kg is travelling on a straight road which is inclined to the horizontal at 5°. At a certain point P on the road the car's speed is $15\,\text{m s}^{-1}$. The point Q is 400 m down the hill from P and at Q the car's speed is $35\,\text{m s}^{-1}$.

(a) Assume that the car's engine produces a constant driving force on the car as it moves down the hill from P to Q, and that any resistance to the car's motion may be neglected. By considering the change in the energy of the car, or otherwise, calculate the magnitude of the driving force of the car's engine.
(b) Assume instead that resistance to the car's motion between P and Q may be represented by a constant force of magnitude 900 N. Given that the acceleration of the car at Q is zero, show that the power of the car's engine at this instant is approximately 12.1 kW.
(c) Given that the power of the car's engine is the same when the car is at P as it is when the car is at Q, calculate the car's acceleration at P.

Q7 A car, of mass 1200 kg, has a maximum power output of 48 000 W. On a horizontal road the car has a maximum speed of $40\,\text{m s}^{-1}$. Assume that the resistance forces acting on the car are proportional to its speed.

(a) Find the resistance force acting on the car when it travels at $v\,\text{m s}^{-1}$.
(b) Find the maximum speed of the car, when it is being driven up a slope at 4° to the horizontal.

Answers can be found on pages 102–103.

12 Calculus in mechanics

Key points to remember

- **Differentiating** the **displacement** gives the **velocity**.

  $v = \frac{ds}{dt}$

- **Differentiating** the **velocity** gives the **acceleration**.

  $a = \frac{dv}{dt} = \frac{d^2s}{dt^2}$

- **Integrating** the **acceleration** gives the **velocity**.

  $v = \int a dt + c$ (Don't forget the constant.)

- **Integrating** the **velocity** gives the **displacement**.

  $s = \int v dt + c$ (Don't forget the constant.)

- In two or three dimensions, differentiate or integrate each component.

  $\mathbf{r} = x\mathbf{i} + y\mathbf{j}$

  $\mathbf{v} = \frac{dx}{dt}\mathbf{i} + \frac{dy}{dt}\mathbf{j}$

  $\mathbf{a} = \frac{d^2x}{dt^2}\mathbf{i} + \frac{d^2y}{dt^2}\mathbf{j}$

Formulae you must know

- $\mathbf{v} = \frac{d\mathbf{r}}{dt}$
- $\mathbf{a} = \frac{d\mathbf{v}}{dt}$
- $\mathbf{v} = \int \mathbf{a} dt + \mathbf{c}$
- $\mathbf{r} = \int \mathbf{v} dt + \mathbf{c}$

Don't make these mistakes ...

- Don't integrate when you should differentiate.
- Don't forget to include a constant when integrating.
- Don't differentiate when you should integrate.
- Don't confuse the rules for differentiating and integrating functions.
- Don't use the formulae for constant acceleration when the acceleration is not constant.

Exam Questions and Answers

How to score full marks

Q1 A particle moves along a straight line, so that its displacement, s metres, from the origin at time t seconds is:

$s = \frac{1}{40}(24t^2 - t^4)$

for $0 \leqslant t \leqslant 2\sqrt{6}$.

(a) Where is the particle when $t = 2\sqrt{6}$?

(b) Find the velocity of the particle when its acceleration is zero.

(a) $s = \frac{1}{40}(24 \times 24 - 576)$

- Substitute $t = 2\sqrt{6}$ into the equation for s.

$= 0$

- Note that $(2\sqrt{6})^2 = 24$.

The particle has returned to the origin.

- Describe the actual position of the particle.

(b) $v = \frac{ds}{dt}$

- Differentiate to find the velocity.

$= \frac{1}{40}(48t - 4t^3)$

$a = \frac{dv}{dt}$

- Differentiate again to find the acceleration.

$= \frac{1}{40}(48 - 12t^2)$

$0 = \frac{1}{40}(48 - 12t^2)$

- Form an equation to find when the acceleration is zero.

$t^2 = 4$

$t = 2$

- Note that $t = -2$ is not within the range of values of t specified in the question, so you need not consider it.

$v = \frac{1}{40}(48t - 4t^3) = \frac{1}{40}(48 \times 2 - 4 \times 2^3)$

- Substitute into the expression for the velocity.

$v = 1.6 \text{ m s}^{-1}$

Q2 A lorry is travelling at 20 m s^{-1} along a straight road, when it begins to brake. While it is braking the acceleration of the lorry is $\left(\frac{t}{10} - 2\right) \text{m s}^{-2}$.

(a) Find the time taken for the lorry to stop.

(b) Find the distance travelled by the lorry as it comes to rest after starting to brake.

(a) $v = \int\left(\frac{t}{10} - 2\right)dt$

- First integrate the acceleration to find an expression for the velocity of the lorry.

$= \frac{t^2}{20} - 2t + c$

- Remember to include a constant of integration.

$v = 20$ when $t = 0 \Rightarrow c = 20$

$v = \frac{t^2}{20} - 2t + 20$

- To find the constant of integration use the initial conditions given: in this case, $v = 20$ when $t = 0$.

When the lorry stops, $v = 0$.

$$0 = \frac{t^2}{20} - 2t + 20$$

$$0 = t^2 - 40t + 400$$

$$0 = (t - 20)^2$$

$$t = 20$$

(b) $s = \int_0^{20} \left(\frac{t^2}{20} - 2t + 20\right) dt$

$$= \left[\frac{t^3}{60} - t^2 + 20t\right]_0^{20}$$

$$= 133\tfrac{1}{3} \text{ metres}$$

Q3 A ship moves so that its velocity, $\mathbf{v}$ m s^{-1}, at time t seconds is $(2 - 0.02t)\mathbf{i} + 4\mathbf{j}$, where $\mathbf{i}$ and $\mathbf{j}$ are unit vectors that are directed east and north respectively. The ship accelerates until it is heading due north; after this time the expression above no longer applies. Take the ship's initial position as the origin.

(a) Find the ship's acceleration.
(b) Find the distance of the ship from the origin when it stops accelerating.

(a) $\underline{a} = \frac{dv}{dt} = -0.02\underline{i}$

(b) When the ship is travelling north:

$$2 - 0.02t = 0$$

$$t = \tfrac{2}{0.02} = 100 \text{ seconds}$$

$$\underline{r} = \int_0^{100} (2 - 0.02t)dt\underline{i} + \int_0^{100} 4dt\underline{j}$$

$$= \left[2t - 0.01t^2\right]_0^{100} \underline{i} + \left[4t\right]_0^{100} \underline{j}$$

$$= 100\underline{i} + 400\underline{j}$$

Distance $= \sqrt{100^2 + 400^2}$
$= 412$ metres

How to score full marks

- To find the displacement of the lorry from its starting position, you need to integrate. In a case like this you can use the times to give the limits of integration.
- Differentiate the velocity to find the acceleration.
- When the ship is travelling north the easterly component of the velocity will be zero. Use this to form an equation that will give the time when the ship is heading north and it stops accelerating.
- To find the displacement of the ship from the origin, you must integrate each component separately. You should use the times $t = 0$ and $t = 100$ as the limits of integration.
- Once you know the displacement, use Pythagoras' theorem to find the distance.

Questions to try

Q1 A moving particle P travels in a straight line. At time t seconds after starting from the point O on the line the velocity of P is $v\,\text{m s}^{-1}$, where $v = t^2(6 - t)$.

Show that the acceleration of P is zero when $t = 4$.

After a certain time, P comes instantaneously to rest at the point A on the line. State the time taken for the motion from O to A, and find the distance OA.

Q2 At time t seconds, a particle P has position vector $\mathbf{r}$ metres relative to a fixed origin O, where $\mathbf{r} = (t^3 - 3t)\mathbf{i} + 4t^2\mathbf{j},\ t \geqslant 0$.

Find:

(a) the velocity of the particle at time t seconds
(b) the time when P is moving parallel to the vector $\mathbf{i} + \mathbf{j}$.

Q3 A particle moves so that its position vector $\mathbf{r}$ at time t is given by:

$$\mathbf{r} = (5t - \frac{t^2}{100})\mathbf{i} + (3t + \frac{t^2}{20})\mathbf{j}$$

where $\mathbf{i}$ and $\mathbf{j}$ are unit vectors directed east and north respectively.

Find the position of the particle when it is travelling due north.

Q4 A particle is initially at the origin. Its initial velocity is $4\mathbf{i}$ and at time t it experiences an acceleration of $t\mathbf{i} + \frac{2t}{5}\mathbf{j}$. Find the time when the position of the particle is $16.5\mathbf{i} + 1.8\mathbf{j}$.

Q5 The acceleration of a particle, of mass 20 kg, at time t seconds is $\mathbf{a}\ \text{m s}^{-2}$, and $\mathbf{a} = \frac{t}{2}\mathbf{i} + 2\mathbf{j}$, where $\mathbf{i}$ and $\mathbf{j}$ are horizontal, perpendicular unit vectors. Initially the particle has velocity $(2\mathbf{i} - 6\mathbf{j})\ \text{m s}^{-1}$ and position $4\mathbf{j}$. Find:

(a) the velocity and position of the particle at time t
(b) the distance of the particle from the origin when $t = 2$
(c) the magnitude of the horizontal force that acts on the particle when $t = 1$.

Q6 As a bird flies between two trees, its position vector at time t seconds is $\mathbf{r}$ m. Two possible models for $\mathbf{r}$, relative to its starting point are proposed.

Model A: $\mathbf{r} = 5t\mathbf{i} + \frac{5}{2}t\mathbf{j}$ Model B: $\mathbf{r} = \frac{5}{2}t^2(3 - t)\mathbf{i} + \frac{5}{4}t^2(3 - t)\mathbf{j}$

where $\mathbf{i}$ and $\mathbf{j}$ are unit vectors directed east and north respectively.

(a) For each model, find the positions of the bird when $t = 0$ and $t = 2$.
(b) For each model, find the velocity of the bird at $t = 0$ and $t = 2$ and explain why the acceleration predicted by model A is zero.
(c) Find the acceleration predicted by model B and the time when it is zero.
(d) Which do you think is the better model? Explain why.

Q7 A particle moves so that at time t seconds its position vector, $\mathbf{r}$ metres, is given by:

$$\mathbf{r} = (4t - 5\sin t)\mathbf{i} + (6t - 3\cos t)\mathbf{j},\ t \geqslant 0$$

where $\mathbf{i}$ and $\mathbf{j}$ are horizontal and vertical unit vectors. Find the time at which the particle is first travelling vertically.

Answers can be found on pages 103–105.

13 Projectiles

Key points to remember

- The solutions to most problems about projectiles start with the **horizontal and vertical displacement** of a projectile.
- Use **constant acceleration equations** to show that $x = Vt\cos\theta$ and $y = Vt\sin\theta - \frac{1}{2}gt^2$.
- The **time of flight** of a projectile is given by the value of t when $y = 0$.
- The **range** is given by the horizontal displacement of the projectile when t is equal to the time of flight.
- At **maximum height**, the vertical component of the velocity of the projectile is zero.
- The **path** of the projectile is given by:
  $y = x\tan\theta - \dfrac{gx^2}{2V^2}(1 + \tan^2\theta)$
- Assume that the **weight** or **force of gravity** is the **only force** that acts on the projectile and there is no air resistance, lift, ...
- A projectile may be modelled as a **particle**, so that its size is not considered.
- Projectiles may land or be launched at different heights and these need to be included in equations.

Formulae you must know

- $x = Vt\cos\theta$
- $y = Vt\sin\theta - \frac{1}{2}gt^2$
- $y = x\tan\theta - \dfrac{gx^2}{2V^2}(1 + \tan^2\theta)$
- $\sin 2\theta = 2\sin\theta\cos\theta$
- $\sec^2\theta = 1 + \tan^2\theta$

Don't make these mistakes ...

Don't ignore the height at which a projectile is launched.

Don't rely on formulae that you have learned for the range, time of flight and maximum height. You should derive these, as you need them, as questions may be set in which these formulae do not apply.

Don't break the motion of the projectile up into a lot of stages. Normally one equation can be used which will apply to the whole motion of the projectile.

Exam Questions and Answers

How to score full marks

Q1 A shot putter throws a shot at an initial speed of $12\,\mathrm{m\,s^{-1}}$, and at an angle of 40° above the horizontal. Assume that the shot is released at a height of 2 m.

(a) Find the range of the shot.
(b) Find the maximum height of the shot.

(a) $-2 = 12\sin 40° t - 4.9t^2$

$0 = 4.9t^2 - 12\sin 40° t - 2$

$$t = \frac{12\sin 40° \pm \sqrt{12^2\sin^2 40° - 4 \times 4.9 \times -2}}{2 \times 4.9}$$

$= -0.2267$ or 1.801

$t = 1.801$ seconds

- First find the time of flight of the projectile, noting that when it lands it will have a displacement of –2 compared to its initial position.

$R = 12\cos 40° \times 1.801$
$= 16.6\,\mathrm{m}$

- Once you have found the time of flight multiply it by the horizontal component of the velocity to find the range.

(b) $0 = 12\sin 40° - 9.8t$

$$t = \frac{12\sin 40°}{9.8}$$

$$H = 12\sin 40° \times \frac{12\sin 40°}{9.8} - 4.9\left(\frac{12\sin 40°}{9.8}\right)^2$$
$= 3.04\,\mathrm{m}$

- The maximum height will be reached when the vertical component of the velocity is zero. Solve this equation to find the time when the projectile is at its maximum height. Then use this height to calculate the actual height.

Q2 A football is kicked with an initial speed of $10\,\mathrm{m\,s^{-1}}$ and at an angle θ above the horizontal. The ball is kicked from the point A and lands at the point B, where both points are on a horizontal football pitch. Given that the distance between the points A and B is 60 m, find the possible values of θ.

$0 = 30\sin\theta t - 4.9t^2$

$$t = 0 \text{ or } t = \frac{30\sin\theta}{4.9}$$

- First find the time of flight for the ball by solving the equation $y = 0$.

$$60 = 30\cos\theta \times \frac{30\sin\theta}{4.9}$$

- Substitute this expression for t into an equation for the horizontal displacement of the ball. Note that you need to use the identity $\sin 2\theta = 2\sin\theta\cos\theta$.

$294 = 900\sin\theta\cos\theta$
$294 = 450\sin 2\theta$
$2\theta = 40.79°$ or $139.21°$
$\theta = 20.4°$ or $69.6°$

- This trigonometric equation has two solutions in the range $0 \leqslant \theta \leqslant 90°$.

Q3 A cannon ball is fired from a cannon at the top of a vertical cliff that has a height of 50 m. The cannon ball hits the water at a distance of 200 m from the base of the cliff. The cannon ball is fired at a speed of $80\,\text{m}\,\text{s}^{-1}$. Find the possible angles of projection of the cannon ball.

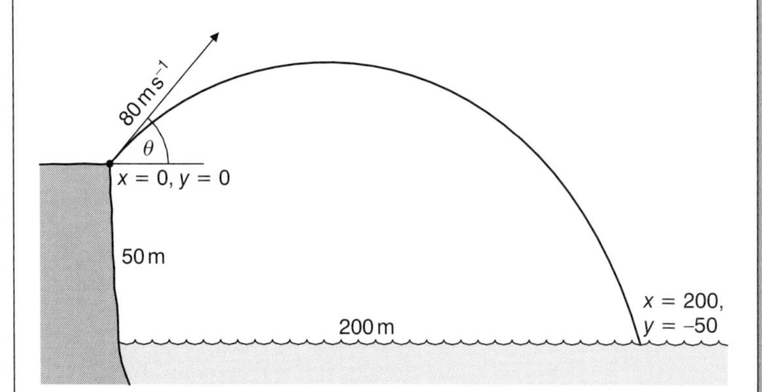

How to score full marks

$200 = 80\cos\theta t$

$t = \dfrac{5}{2\cos\theta}$

$-50 = 80\sin\theta t - 4.9t^2$

$-50 = 80\sin\theta \times \dfrac{5}{2\cos\theta} - 4.9\left(\dfrac{5}{2\cos\theta}\right)^2$

$-50 = 200\tan\theta - 30.625\sec^2\theta$

$-400 = 1600\tan\theta - 245\sec^2\theta$

$-400 = 1600\tan\theta - 245(1 + \tan^2\theta)$

$0 = 245\tan^2\theta - 1600\tan\theta - 155$

$\tan\theta = \dfrac{1600 \pm \sqrt{1600^2 - 4 \times 245 \times -155}}{2 \times 245}$

$= -0.09548$ or 6.626

$\theta = -5.5°$ or $81.4°$

The angle is either 5.5° below the horizontal or 81.4° above the horizontal.

- Use the horizontal displacement to express t in terms of θ.
- Substitute this expression for t into an equation for the vertical motion and simplify, using the identity $\sec^2\theta = 1 + \tan^2\theta$ to give a quadratic equation in $\tan\theta$.
- This quadratic can be solved to give two values of $\tan\theta$ and hence two values of θ.
- Note that the negative θ indicates that the angle is below the horizontal.

Questions to try

Q1 A cricket ball is hit from a height of 0.8 m above horizontal ground with a speed of $26\,\text{m s}^{-1}$ at an angle α above the horizontal, where $\tan\alpha = \frac{5}{12}$. The motion of the ball is modelled as that of a particle moving freely under gravity.

(a) Find, to 2 significant figures, the greatest height above the ground reached by the ball.

When the ball has travelled a horizontal distance of 36 m, it hits a window.

(b) Find, to 2 significant figures, the height of the ball when it hits the window.
(c) State one physical factor that could be taken into account in any refinement of the model which would make it more realistic.

Q2 During a practice session, a basketball player throws a ball towards a horizontal ring of centre A. In a simple model this ball is treated as a particle.

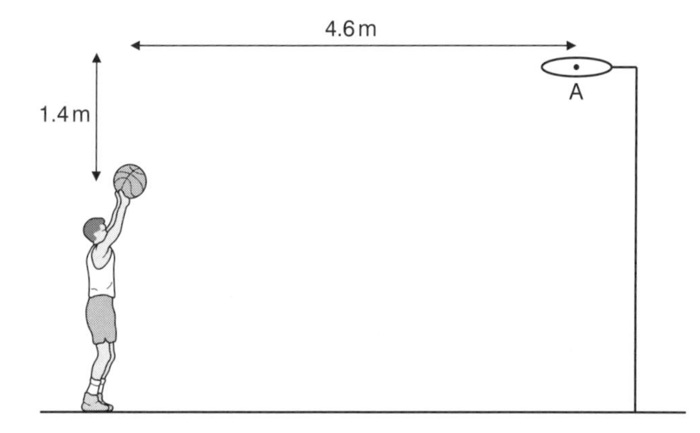

(a) The ball is projected at a speed of $8\,\text{m s}^{-1}$ and at an angle of 40° to the horizontal, from a point at a horizontal distance of 4.6 m from A and 1.4 m below A.
  (i) Find the time taken for the ball to travel a horizontal distance of 4.6 m, giving your answer correct to 2 significant figures.
  (ii) Taking $g = 9.8\,\text{m s}^{-2}$, show that the ball passes below A.

(b) The player then throws again from the same point as before. He projects the ball at an angle of 40° to the horizontal but increases the speed of projection to $V\,\text{m s}^{-1}$.
  (i) Determine the value of V for which the ball passes through A.
  (ii) Show that for this value of V the ball is descending as it passes through A.

Q3 An archer fires an arrow at a speed of $28\,\text{m s}^{-1}$ and at an angle θ above the horizontal. The arrow is modelled as a particle that is initially at a height of 1.5 m. The arrow hits a target at a height of 2 m. The target is a horizontal distance of 40 m from the point where the arrow is fired.

(a) Find the possible values of θ, correct to the nearest degree.
(b) Which angle do you think was most likely to have been used? Explain why.

Questions to try

 A shell is fired from a stationary ship O which is at a distance of 1000 m from the foot of a vertical cliff AB of height 100 m. The shell passes vertically above B and lands at a point C on horizontal ground, level with the top of the cliff (see diagram).

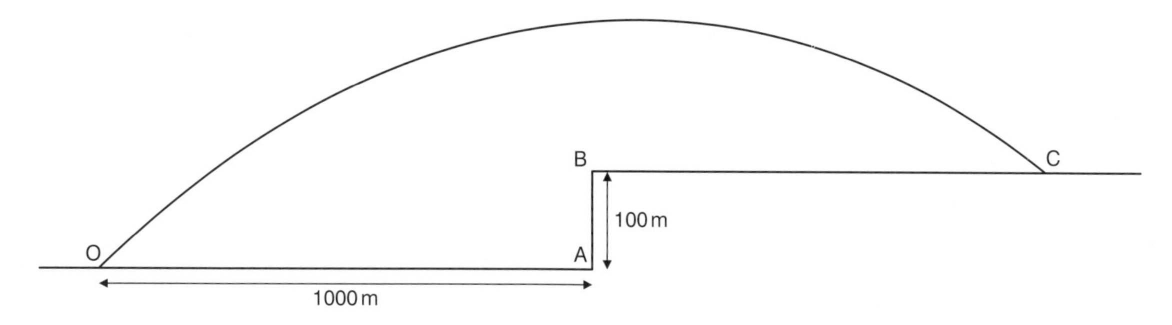

The shell is fired with speed 300 m s^{-1} at angle of elevation θ and air resistance to the motion of the shell may be neglected.

(a) Given that $\theta = 30°$, find the time of flight of the shell and the distance BC.
(b) Given instead that the shell *just* passes over B, as shown in the second diagram, find the value of θ, correct to the nearest degree.

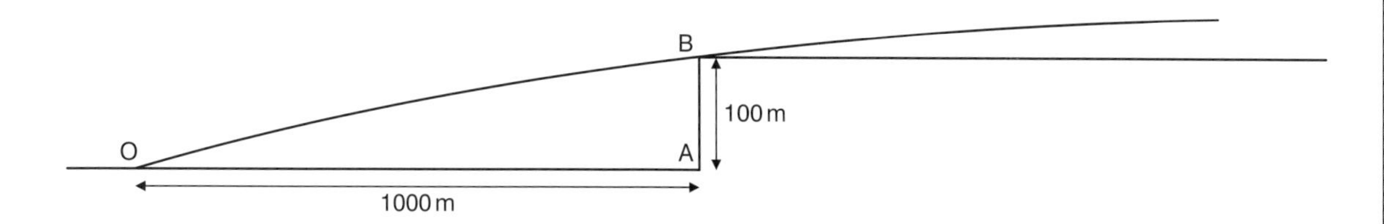

 A golfer hits a ball from a point O, with an initial velocity of 28 m s^{-1} at an angle θ above the horizontal. The ball clears an electricity pylon. The height of the pylon is 27.5 m and is at a horizontal distance of 20 m from O.

(a) State one assumption that it is important to make about the motion of the ball in order to model the motion of the ball.
(b) Find the possible range of values of θ.
(c) Find an expression for the range of the ball in terms of θ.
(d) If the ground is horizontal explain which value of θ gives:
  (i) the shorter time of flight
  (ii) the greater range.

Answers can be found on pages 105–106.

14 Momentum and the coefficient of restitution

Key points to remember

- Use v for velocity after collision, u for velocity before collision, e for the **coefficient of restitution**.
- For a collision with a **wall or barrier**, $v = -eu$. Note that the sign of the velocity changes.
- For a collision between **two bodies**, use:
  - $v_A - v_B = -e(u_A - u_B)$

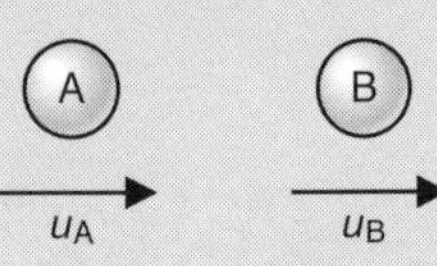
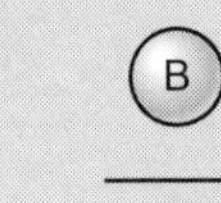

  - **conservation of momentum**.
- For a perfectly **elastic collision** $e = 1$.
- For an **inelastic collision** $e = 0$.
- Revise **impulse** from the AS module.

Formulae you must know

- For a perpendicular collision with a wall: $v = -eu$
- For a collision between two bodies: $v_A - v_B = -e(u_A - u_B)$
- Conservation of momentum: $m_A u_A + m_B u_B = m_A v_A + m_B v_B$
- Impulse: $I = Ft = mv - mu$

Don't make these mistakes ...

Don't miss out minus signs when dealing with velocities in the negative direction. This is probably the most common error in this topic.

Don't confuse initial and final velocities, especially if more than one collision is involved.

Exam Questions and Answers

How to score full marks

Q1 A ball of mass 250 grams is dropped onto a horizontal surface, which it hits when it is moving at $6\,\text{m s}^{-1}$. Calculate the magnitude of the impulse on the ball. The coefficient of restitution between the ball and the surface is 0.6.

$v = 6$ ↓ ○ ↑ v ↓ positive direction

$v = -0.6 \times 6$
$= -3.6\,\text{m s}^{-1}$

- Use $v = -eu$ to calculate the rebound velocity.

$I = 0.25 \times -3.6 - 0.25 \times 6$
$= -2.4\,\text{Ns}$

- Use $I = mv - mu$ to find the impulse, making sure that you use the correct sign with each velocity.

Magnitude of the impulse = 2.4 Ns

- You do not need the minus sign for the magnitude.

Q2 A sphere A has mass 2 kg and is moving along a straight line on a smooth surface at $5\,\text{m s}^{-1}$. It collides with the sphere B, which has mass 3 kg and is moving along the same straight line towards A at $3\,\text{m s}^{-1}$. The coefficient of restitution between the two spheres is 0.4. Describe what happens to the two spheres after they collide.

before: (A) → $v_A = 5$ + → after (B) ← $v_B = -3$

before: (A) → v_A after (B) → v_B

- Draw a diagram showing the velocities before and after the collision.

By conservation of momentum:

$2 \times 5 + 3 \times -3 = 2v_A + 3v_B$
$1 = 2v_A + 3v_B$

- First use conservation of momentum to form an equation relating the two final velocities.

Using the coefficient of restitution:

$v_A - v_B = -0.4(5 - (-3))$
$= -3.2$
$v_A = v_B - 3.2$

- Use the coefficient of restitution to form a second equation. Be very careful with the signs of the velocities when forming these equations. Solve these equations to find the velocities after the collision.

Substituting:

$1 = 2(v_B - 3.2) + 3v_B$

$1 = 5v_B - 6.4$

$v_B = 7.4 \div 5 = 1.48$

$v_A = 1.48 - 3.2 = -1.72$

Both spheres change direction so that they are moving away from each other, A moves at $1.72\,\text{m s}^{-1}$ and B moves at $3.2\,\text{m s}^{-1}$.

How to score full marks

- Give a description to show that you are able to explain the situation after the collision.

Questions to try

Q1 A small ball of mass 0.25 kg is dropped from a height of 1.2 m above a horizontal floor. The ball rebounds vertically from the floor, reaching a height of 0.8 m. Assuming that air resistance can be neglected, calculate:

(a) the coefficient of restitution between the ball and the floor

(b) the impulse exerted by the floor on the ball when it bounces.

If air resistance were taken into account, would the value calculated for the coefficient of restitution be larger or smaller than the value calculated in part (b)? Justify your conclusion.

Q2 A smooth sphere A moves in a straight line across a smooth horizontal surface with speed U. The sphere A moves directly towards a stationary smooth sphere B, as shown in the diagram, and subsequently A and B collide. The spheres A and B are of equal radius and their masses are $3m$ and m respectively.

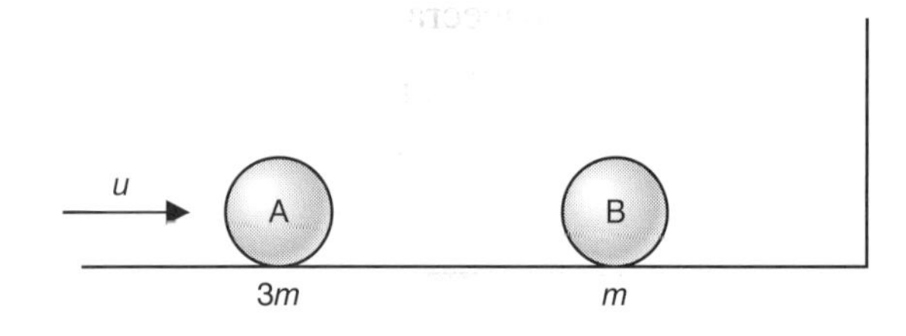

(a) Assuming that the collision between the spheres is perfectly elastic, show that after the collision the sphere B has speed $\frac{3U}{2}$.

(b) The sphere B subsequently collides with a wall which is perpendicular to its direction of motion. The coefficient of restitution of restitution between the sphere B and the wall is e. Given that after B rebounds from the wall, both spheres A and B are moving with the same speed, find the value of e.

Q3 Two particles P and Q are travelling directly towards each other along a straight line. Both particles have the same speed of $3\,\text{m s}^{-1}$. After they have collided the particle Q is brought to rest and the direction of motion of the other particle is reversed. Given that the mass of Q is twice the mass of P, find the speed of P after the collision and the coefficient of restitution between the two particles.

Questions to try

A particle A of mass m is moving with speed $3u$ on a smooth horizontal table when it collides directly with a particle B of mass $2m$ which is moving in the opposite direction with speed u. The direction of motion of A is reversed by the collision. The coefficient of restitution between A and B is e.

(a) Show that the speed of B immediately after the collision is $\frac{1}{3}(1 + 4e)u$.
(b) Show that $e > \frac{1}{8}$.

Subsequently B hits a wall fixed at right angles to the line of motion of A and B. The coefficient of restitution between B and the wall is $\frac{1}{2}$. After B rebounds from the wall, there is a further collision between A and B.

(c) Show that $e < \frac{1}{4}$.

Two spacecraft, A and B, with masses m and M respectively, are moving along the same straight line during an attempt to link up. There is a direct collision which occurs when the two spacecraft are approaching each other with speed u, as shown in the diagram. The spacecraft fail to link up and after the collision each moves in the opposite direction to its original motion. Spacecraft A now has a speed of $\frac{u}{2}$.

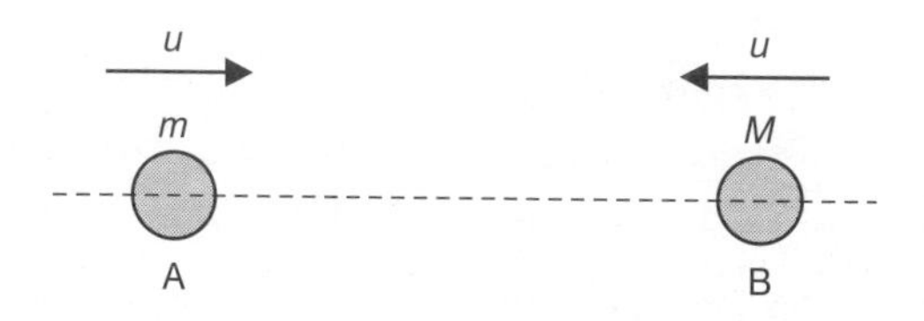

(a) Draw diagrams to show the velocities of A and B before and after the collision. Show that after the collision the speed, v of B is $u\left(\frac{3m}{2M} - 1\right)$ and explain why $\frac{m}{M} > \frac{2}{3}$.

(b) The coefficient of restitution between the spacecraft in the collision is e. Use the fact that $e \leqslant 1$ to show that $v \leqslant \frac{3u}{2}$ and hence $\frac{m}{M} \leqslant \frac{5}{3}$.

The rocket motor of spacecraft B is now fired so that its direction of motion is reversed. It catches up spacecraft A and links up with it. You may assume that there is a negligible loss of mass from spacecraft B when the rocket motor is used.

(c) If the combined spacecraft has a speed of $\frac{3u}{4}$, calculate the impulse given to B by its rocket motor in terms of m and M.

Answers can be found on pages 107–108.

15 The Poisson distribution

Key points to remember

- The Poisson distribution is used for events that occur **randomly in a continuous medium** (i.e. space or time). For example emergency calls to an ambulance service may be considered to be random events, and they occur in the medium of time.
- The Poisson distribution is used to model the number of these events that occur in a **fixed interval** of space or time, for example the number of emergency calls in a period of 24 hours.
- These events must be **independent**, they must occur **singly**, their **occurrence must be uniform**. This means that no two emergency calls could be received at the same moment in time, and that there are generally no particularly busy or slack times.
- The **rate of occurrence** is often represented by the Greek letter λ. This is the average number of events in a fixed interval, for example it is the average number of emergency calls in 24 hours.
- The **probability** of there being ***r* events** is given by the expression:
  $\dfrac{e^{-\lambda} \times \lambda^r}{r!}$
- The **mean number** of occurrences is λ, and the **variance** of the number of occurrences has the same value. This leads to the **standard deviation** of the number of occurrences being $\sqrt{\lambda}$.
- Independent Poisson distributions are **additive**. If the mean number of emergency calls at one ambulance centre is λ_1 and the mean at another centre is λ_2 then the distribution for the total number of calls is Poisson with mean equal to $\lambda_1 + \lambda_2$.
- Tables of the Poisson distribution show the probabilities for *r* or fewer events in the fixed interval when the mean number is λ. You will not find all possible values of λ listed.
- Some calculators will calculate Poisson probabilities or cumulative probabilities directly.
- The Poisson also serves as a model for the occurrence of **rare events**. This could be where a binomial distribution is appropriate for a large number of events with a very small probability of success.

Formulae you must know

- $P(R = r) = \dfrac{e^{-\lambda} \times \lambda^r}{r!}$
- $0! = 1$ and $\lambda^0 = 1$, hence $P(R = 0)$ is $e^{-\lambda}$.

Don't assume that the Poisson is always to be used for random events. Independence, singularity and uniformity must hold.

Exam Questions and Answers

How to score full marks

Q1 The number of accidents that occur on a stretch of road, in a four-week period, is modelled using a Poisson distribution with mean 2.1.

(a) What is the probability of exactly four accidents occurring in four weeks?
(b) Find the probability of at least four accidents occurring in this period.
(c) What is the probability of at least eight accidents occurring in eight weeks?

(a) Here $\lambda = 2.1$ and so $P(R = 4)$

$$= \frac{e^{-2.1} \times 2.1^4}{4!} = 0.0992.$$

- The question tells you that the Poisson distribution is to be used, so do not do anything else.
- Use your calculator for this sort of calculation, or use tables of the distribution function: $P(\leqslant 4) - P(\leqslant 3)$.

(b) $P(R < 4) = P(0) + P(1) + P(2) + P(3)$
$= 0.8386$
$P(\text{at least } 4) = 1 - P(< 4) = 0.1614$

- Don't make this difficult, when distribution tables make it straightforward. Don't risk summing separate Poisson probabilities yourself.

(c) Now $\lambda = 4.2$, hence $P(R < 8)$ is 0.9361.

$P(\text{at least } 8) = P(\geqslant 8) = 1 - P(< 8)$
$= 1 - 0.9361 = 0.0639$

- This is about the additive property of the Poisson distribution.
- Don't make the mistake of thinking that everything just gets doubled.

Q2 A telephone sales worker makes 40 calls per hour. On average, he is successful in converting two of these into actual contracts. Let X be the number of sales he makes in an hour's work. Find, to an accuracy of three decimal places:

(a) $P(X = 2)$
(b) $P(X \text{ is at least } 6)$

(a) $\lambda = 2$, hence $P(X = 2)$ is 0.2707
$= 0.271$ (3 d.p.)

- Strictly this is a situation where the binomial B(40, 0.05) is the more appropriate probability model to use, but it can be treated as a rare events situation.
- Don't throw away marks by ignoring instructions on the degree of accuracy required.

(b) $P(X \geqslant 6) = 1 - P(X \leqslant 5)$,
hence $1 - 0.9834 = 0.017$ (3 d.p.)

- Remember that 'at least' is the same as $\geqslant$.
- A good way to tackle this open-ended event is to use its complement.

How to score full marks

Q3 Which probability model would you use to tackle the following problems?

(a) $P(X < 5)$, given that $X \sim B(100, 0.012)$
(b) $P(X < 5)$, given that $X \sim B(10, 0.12)$

(a) $Poi(\lambda = 1.2)$

- This is the 'rare event situation', a binomial context where the probability (0.012) is very small and the mean (1.2) is also small.

(b) Binomial with $n = 10$ and $p = 0.12$

- This one is a fairly standard binomial situation and needs no approximations.
- Neither part asks you to find the probability, so don't waste your time trying to.

Q4 A fire-officer notes that one watch in a hundred is completely incident free. (A watch is an 8-hour continuous period of duty for firefighters.) What is the average number of incidents per watch?

$P(R = 0)$ is $e^{-\lambda}$, hence $\lambda = -\ln 0.01 = 4.6$

- Use the few clues provided: continuous period, incident, fire-officer.
- Make sure you are up to speed with your pure maths and exponential functions. Inverting the exponential function involves the natural logarithm.

Questions to try

Q1 A manufacturer finds that, overall, 1.5 per cent of computers they supply develop a fault within one month of being sold. A college buys 200 computers from this manufacturer. What is the probability that fewer than five will develop a fault within one month? State clearly the assumptions you are making to justify your choice of probability distribution.

Q2 In a northern laboratory, radioactive events occur randomly at an average rate of six per minute. In a southern laboratory, the likelihood of there being no radioactive event in one second is 0.606 53.

(a) Find the probability that there are at least three radioactive events in one minute in the northern laboratory.
(b) Find the probability that there are at least three events in six seconds in the southern laboratory.
(c) Find the probability that there are exactly six events in total in the two laboratories in a period of fifteen seconds.

Q3 State the conditions under which a binomial distribution $B(n, p)$ may reasonably be approximated by a Poisson distribution $Poi(\lambda)$. Ensure that you show clearly the relationship between the parameters.

The probability that a given person has blood of type AB– is assumed to be 0.014. There are 300 students in the sixth form of a community college. It is suggested that a binomial probability distribution is appropriate to determine an estimate of the probability that at least five students have blood of type AB–.

(a) Under what circumstances would this be an appropriate choice of probability model?
(b) Determine the probability of this event and justify the probability model you use.

Answers can be found on page 109.

16 The chi-squared (χ^2) distribution

Key points to remember

- The chi-squared distribution, χ^2, is used with **frequencies**.
- It can be used to test whether a set of observed frequencies shows evidence of **independence** or the opposite, **association**.
- This use of χ^2 is labelled 'contingency table'.
- The observed frequencies are compared with those expected by the independence model.
- Expected frequencies should not be less than 5.
- The **null hypothesis** is that the factors are independent. The **alternative hypothesis** is that the factors are not independent.
- The **independence model** in probability implies that the probability that *A* and *B* occur together is found directly by multiplying the probability for *A* by the probability for *B*.
- The **expected frequencies** are found by multiplying row and column sub-totals and dividing by the grand total.
- The **relative difference** between the **observed** and the **expected** frequencies is $\frac{(O-E)^2}{E}$.
- The sum of these relative differences provides the **test statistic**.
- A large value for the test statistic corresponds to a large difference between expected and observed frequencies, and implies poor fit.
- **Degrees of freedom** are an important consideration and play a major role in using chi-squared.
- In a contingency table with *r* rows and *c* columns, there are $(r-1)(c-1)$ degrees of freedom e.g. a table with two rows and four columns has $(2-1)(4-1) = 1 \times 3 = 3$ degrees of freedom.
- A 2×2 table has just one degree of freedom and **Yates' correction** is an important consideration.
- Some important critical values at five per cent are 3.841, 5.991, 7.815 and 9.488 with one, two, three and four degrees of freedom respectively.
- Very small values of the test statistic suggest a suspiciously close fit between expected and observed frequencies; in these circumstances the **left-tail critical values** should be used.
- The goodness-of-fit of other probability models, such as binomial, Poisson or normal, may be tested using the χ^2 distribution.
- The number of degrees of freedom is determined by the number of **frequency cells** and the number of **constraints** on the model, such as the total of the expected frequencies and, in the binomial case, the probability, the mean in the case of the Poisson, and either or both mean and variance in the case of the normal.

Formulae you must know

- In a contingency table: $E = \frac{R \times C}{N}$
- Degrees of freedom: $\nu = (r-1)(c-1)$
- Test statistic $= \sum \frac{(O-E)^2}{E}$
- Yates' correction is: $\sum \left(\frac{(|O-E| - 0.5)^2}{E} \right)$

Don't make these mistakes ...

- When finding the test statistic, don't divide by the grand total.
- Don't misinterpret the outcome of the test: large values of the test statistic imply big differences and this implies poor fit.
- Always check your number of degrees of freedom.
- When using Yates' correction, find the absolute difference then subtract a half **before** squaring.
- Don't get the right and left tails confused.

Exam Questions and Answers

Q1 Are mode of travel to school and gender of child associated? Answer this question in the light of the evidence provided by one hundred children, as shown in the table below. Conduct your test at the five per cent level of significance.

	By car	On foot
Boy	44	19
Girl	24	13

The expected frequencies are:

42.84	20.16
25.16	11.84

$\nu = 1$, therefore the critical value is 3.841.

$\chi^2 = 0.086$ which is less than 3.841, therefore the result is not significant and so there is no evidence of association between gender and mode of transport.

- You need to use Yates' correction. The value of the statistic is: $\frac{(|44 - 42.84| - 0.5)^2}{42.84} + \ldots$
- This question gives very little in the way of guidance, so you need to demonstrate all you know that is relevant.
- Always show your E values, even if the question does not ask for them.
- Be consistent in giving your E values.
- Check the row and column totals for your E value.
- Give the critical value, and the number of degrees of freedom.
- You do not need to write down every calculation, but you do need to show you understand the question.
- Make sure you give the final value of the test statistic.
- If you use a calculator to perform this automatically, be careful – they do not necessarily use Yates' correction.
- State your conclusions with care and precision.

How to score full marks

Q2 In a survey on smoking, 200 people at a college were asked if they were in favour of allowing smoking on campus. The table shows the results.

	Yes	Don't know	No
Teachers	28	42	10
Other staff	9	11	40
Students	26	24	10

(a) Assuming that respondent and opinion are independent, determine the number expected in each category.
(b) Use an appropriate test to find whether the respondent's role and response are independent.
(c) Report your findings fully.

(a) Expected frequencies are:

25.2	30.8	24.0
18.9	23.1	18.0
18.9	23.1	18.0

(b) H_0: independent, H_1: not independent, degrees of freedom = 4, critical value = 9.488, the test statistic is $((28 - 25.2)^2 \div 25.2 + \ldots) = 57.221$

(c) This is a highly significant result. The opinion and role of the respondent are linked.

- Independence is the basis of all contingency table tests.
- Record the expected frequencies, correct to at least one decimal place.
- Mention the hypotheses, degree of freedom, and critical value.
- Show more calculation than here.
- Let your calculator do the arithmetic.
- You might also want to say that the 'other staff' are very much less likely to be against smoking on campus, and the teachers are of the opposite opinion.

Q3 Two coins were thrown 20 times and the number of 'heads' scored was recorded each time. The outcomes are shown in the table.

No. heads	0	1	2
Frequency	4	11	5

A binomial distribution B(2, 0.5) is used to model these outcomes. Test at the five per cent level of significance if this model is appropriate.

E values: $20 \times 0.5^2 = 5$,
and $20 \times 2 \times 0.5^2 = 10, \ldots$

$\chi^2 = \frac{1}{5} + \frac{1}{10} + 0 = 0.3$

The critical value is 5.99 so, not significant. The model is a good fit for the data.

- The question does not ask for hypotheses but it would be wise to state them, e.g. H_0: binomial is appropriate, H_1: not appropriate.
- Also mention degrees of freedom.

Questions to try

Q1 The blood type of each person in 300 families with five family members was recorded. The table shows the number of people with type-O blood.

No. people with type-O	0	1	2	3	4	5
No. families	18	81	85	70	38	8

(a) Assume that a binomial distribution with $p = 0.45$ is appropriate to model these data. Determine the expected frequencies.
(b) Test whether this model is appropriate.

Q2 The English teacher in a small rural school gave 55 children in year 7 a piece of written work. The numbers of errors made by the children are recorded in the table below.

No. errors	0	1	2	3	4	5	6	7	$\geqslant 8$
No. children	4	13	13	14	7	2	1	1	0

(a) Use the data to determine the parameter of the Poisson distribution which might model the number of errors made by a year 7 child.
(b) Test the hypothesis of the appropriateness of the Poisson distribution to model the frequencies observed.
(c) State your conclusions clearly.

Q3 The following table shows the number of books sold according to the type of cover (hardback or paperback) and the nature of the book (reference, light reading, serious reading).

	Reference	**Light**	**Serious**
Paperback	12	66	32
Hardback	42	8	38

Use a χ^2 test at the five per cent level to determine whether there is evidence of association between a book's cover and its content.

Q4 The numbers of convictions for different types of car crimes and the ages of those convicted are recorded in the table.

Age	**Theft**	**Licence**
Under 20	23	25
20 and over	17	35

Use a χ^2 test (including Yates' correction) at the five per cent level of significance, to test whether type of car crime is independent of age of the criminal.

Answers can be found on page 110.

17 Hypothesis testing

Key points to remember

- The **null hypothesis** is used to build the probability model on which the statistical test is based.
- The **alternative hypothesis** represents the position adopted when the null hypothesis is rejected.
- **Statistical tests** may be one-tailed or two-tailed, according to the nature of the alternative hypothesis.
- A **critical value** of the test statistic defines a boundary between acceptance and rejection of the null hypothesis.
- Results that fall within the **critical region** lead to rejection of the null hypothesis.
- There is a small but non-zero likelihood of a result falling in the critical region when the null hypothesis holds, the probability of this happening is the **significance** (or **alpha**) level of the test.
- Two possible errors may occur as the result of otherwise appropriately conducting an hypothesis test:
  (a) rejecting the null hypothesis when it actually holds – this is a **Type I** error
  (b) accepting the null hypothesis when it does not hold – this is a **Type II** error.
- The chance of a Type I error is the **significance level**. The chance of a Type II error is more difficult to determine.

Formulae you must know

- H_0 and H_1 are often used for the null and alternative hypotheses.
- α is often used for the significance level.
- In correlation tests, H_0 is 'not correlated'.
- The standard error of the mean is: $\frac{\sigma}{\sqrt{n}}$
- In *z*-tests, the test statistic is: $\frac{\bar{x} - \mu}{\sigma/\sqrt{n}}$
- In tests of sample proportion *p* is approximately $N(p, \frac{pq}{n})$)

Don't make these mistakes ...

Exam Questions and Answers

Q1 An ordinary coin is to be tested for bias.

(a) What are appropriate null and alternative hypotheses?

(b) The coin was thrown 100 times. It landed tails 56 times. What conclusions do you draw?

(a) H_0: $p = 0.5$, H_1: $p \neq 0.5$.

(b) $p \sim N(p, \frac{pq}{n})$, so the test statistic is:

$$\frac{0.56 - 0.5}{\sqrt{\frac{0.5 \times 0.5}{100}}}$$

which is 1.2 and is much less than the five per cent level of 1.96. So the coin is not biased.

Q2 The following data were extracted from a daily newspaper over a period of eight days in October.

Rainfall, x cm	1.3	3.8	4.2	2.6	2.1	2.6	5.3	0.9
Sunshine, y hours	1.5	0.3	0.0	4.2	3.6	0.5	0.0	1.4

A test is to be carried out to determine whether there is evidence of correlation in these data.

(a) State the appropriate statistic to calculate, giving your reasons.

(b) Calculate the value of the statistic.

(c) Make your conclusions clear.

(a) Spearman's rank, since a straight-line correlation is not sought.

(b) Ranks are:

7	3	2	4.5	6	4.5	1	8
3	6	7.5	1	2	5	7.5	4

and r_s comes to –0.7108.

(c) The five per cent critical value is –0.6429, hence there is evidence of negative correlation i.e. as sunshine decreases so rainfall increases.

How to score full marks

- You would expect half the outcomes to be tails, so this is the null hypothesis.
- The alternative hypothesis does not stipulate the nature of bias in this case.
- This is a test of proportion.
- Show your working.
- Evaluate relevant expressions.
- Quote critical value and significance level.
- Make you conclusion clear.

- There are clues here in the word 'correlation' and the fact that there is no indication of a need to establish linearity.

- It might be better to give a fuller statement and to include '… correlation coefficient'.
- You need to show the ranks since the figures are provided.
- There are ties, so use the product moment correlation coefficient for the ranks (the Σd^2 formula gives –0.6905).
- Note that the answer gives both the significance level and the critical value.
- It would be wise to state the hypothesis.

Q3 It is established that the distribution of heights of a particular species of plant is normal with standard deviation 9 cm. The mean height of a random sample of 20 plants is 45.2 cm.

(a) Test, at the five per cent level of significance, the hypothesis that the mean height of the population is less than 49 cm.
(b) Explain briefly what is meant, in the context of this question, by a Type I error.
(c) Explain briefly what is meant, in the context of this question, by a Type II error, and calculate the probability of making a Type II error when $\mu = 40$ cm.

(a) Test statistic is:

$$\frac{45.2 - 49}{\frac{9}{\sqrt{20}}} = -1.888$$

and the critical value is −1.645. Hence the population mean could be less than 49 cm.

(b) The null hypothesis $\mu = 49$ would be rejected when the mean height is 49 cm.

(c) The null hypothesis $\mu = 49$ would be accepted when in fact the mean height is less than this.

This is when $\bar{x} > 49 - 1.96 \times \frac{9}{\sqrt{20}}$ $= 45.06$ and $\mu = 40$.

So the probability is

$$1 - \Phi\frac{(45.06 - 40)}{9/\sqrt{20}} = 0.00596$$

Q4 In the Basil Hotel, the probability of a guest being dissatisfied with the service is 0.25. When a random sample of 20 guests were contacted a week after their stay, three expressed dissatisfaction.

State your hypotheses clearly and test, at the five per cent level of significance, whether the proportion of dissatisfied customers is unusually low.

H_0: $p = 0.25$, H_1: $p < 0.25$

B(20, 0.25): $P(x \leqslant 3) = 0.225$ which is not significant at five per cent.

How to score full marks

- Make it clear that you are calculating the statistic correctly and evaluate the expression.
- Give the critical value.
- Make your conclusion clear.
- Give your answer, using the setting of the question.
- Give a clear derivation of the appropriate value.
- A diagram may help you sort out the details – so use one if you are in any doubt at all.
- You should realise that you are in a position to have gleaned the answer to this from the information provided, up to this point.
- The one-sided test is indicated by 'low'.
- Make sure you know the conditions for various probability models.

Questions to try

Q1 A manufacturer finds that, overall, 35 per cent of computers they supply develop a fault within one month of being sold. A school buys 20 computers from this manufacturer. Four develop a fault within one month. State clearly the hypotheses you are using and test, at the five per cent level, the claim that these computers are more reliable.

Q2 In a southern laboratory, radioactive events occur randomly at an average rate of eight events per minute. A chemist records the background radiation in her own laboratory with the result that there are 14 events per minute. Test the claim that the count is higher in the chemist's laboratory.

Q3 It is felt that the strength of a certain manufacturer's twine is normally distributed with a mean of 8.5 kN and a standard deviation of 0.51 kN. A sample of ten pieces of twine is tested. The mean breaking strength of the sample is 8.02 kN.

(a) Test the hypothesis that, at the 2.5 per cent significance level, the mean of the population is less than 8.5 kN.

(b) Describe, in the context of this question, a Type I error and state the probability of making such an error.

Q4 During the lambing season, eight ewes and the lambs they bore were weighed at the time of birth.

Ewe	A	B	C	D	E	F	G	H
Weight of ewe, x kg	49	46	48	45	46	42	43	40
Weight of lamb, y kg	3.7	3.0	3.4	2.9	3.1	2.7	3.0	2.8

Test, at the 2.5 per cent level of significance, the claim that there is a positive linear relation between a ewe's weight and that of its lamb.

Q5 Forestry officials claim that trees grow more slowly, the greater the altitude at which they are planted. The growth rates of seven otherwise identical trees planted at different altitudes were measured. Altitude A is less than B and so on, growth rate 1 was the fastest.

Altitude	A	B	C	D	E	F	G
Growth rate	1	4	3	5	6	7	2

Test the claim , at the five per cent level of significance.

Q6 A new drug for the common cold is trialled. The manufacturers claim that it is at least 75 per cent effective within one week of commencing the course of treatment. A group of 40 volunteers each commence taking the drug once they contract a cold. Of these, 36 report being cured within one week.

(a) State appropriate hypotheses for this test.

(b) The test is to be conducted at the two per cent level of significance. What is the critical region?

(c) Carry out the test and report your findings fully.

Answers can be found on page 111.

18 Approximating distributions

Key points to remember

- Approximations should be used only when appropriate.
- Using a continuous random variate to approximate a **discrete** random variate involves a **continuity correction**.
- Approximating a **binomial distribution** $B(n, p)$ using the **Poisson distribution** requires p to be small and n to be large.
- The closer that p in the **binomial distribution**, $B(n, p)$, is to 0.5 and the larger that n is, the better the **normal approximation**, $N(np, np(1-p))$, is.
- Approximating a **Poisson distribution** $Poi(\lambda)$ using the **normal distribution** requires λ to be large.
- Probabilities associated with sums of n **independent** observations from the same distribution may be approximated using the **normal distribution** $N(n\mu, n\sigma^2)$.

Formulae you must know

- Standard deviation is $\sqrt{\text{variance}}$.
- In $B(n, p)$, the mean is np and the variance is $np(1-p)$.
- In $Poi(\lambda)$, the mean is λ and the variance is λ.
- The continuity correction involves $x \pm \frac{1}{2}$, rather than x.

Don't make these mistakes ...

Go directly to the approximation – **don't approximate an approximation**. For example, do not approximate a binomial by a Poisson, then decide that the mean of the Poisson is large enough to approximate the Poisson by a normal.

Don't use the continuity correction when you are approximating a continuous random variable.

Exam Questions and Answers

Q1 An environmentalist is observing the behaviour of cattle in a pasture. The pasture has been sectioned into a number of identical regions. There are, on average, 1.05 cattle per region.

- This is important.

(a) Assuming that the cattle are randomly scattered in the pasture, describe an appropriate probability model for the number of cattle per region.

- 'Randomly scattered' is the key phrase here.

(b) On another farm the average number of cattle per region is 16. Calculate, using a suitable approximation, the probability that a randomly selected region contains fewer than 12 cattle.

(a) Poisson with mean 1.05.

- The description needs only to be the key parameters.

(b) Poi(16) ~ N(16, 16).

- Make the approximation clear.

$P(X < 12)$ in Poisson is approximated by $P(\leqslant 11.5)$ in the normal.

- Continuity correction clearly attempted.
- An appropriate diagram could be valuable.

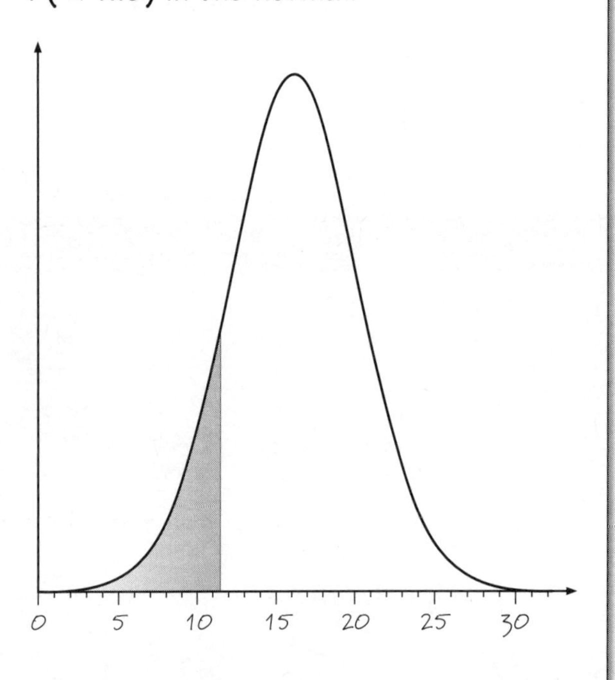

Hence $1 - \Phi\left(\frac{16 - 11.5}{\sqrt{16}}\right)$

$= 0.3697$ i.e. 0.370

- This shows how the approximate probability is being evaluated.

How to score full marks

Q2 On average, three per cent of the output of an assembly process is faulty. A random sample of 50 items is obtained. The number of faulty items is X.

(a) State the appropriate probability distribution for X and give its parameters.
(b) Calculate the exact probability that $X = 1$.
(c) Explain why it may be appropriate to use a Poisson model to approximate the distribution of X. Use this model to determine the probability of finding at least two faulty items.

(a) $X \sim B(50, 0.03)$

- Clarify that you are using the binomial probabilities here.

(b) $P(X = 1) = 50 \times 0.03 \times 0.97^{49}$
$= 0.3372$

(c) The binomial has a small value of p and a large value of n.

- Make the conditions clear.

$P(X \geqslant 2) = 1 - (P(0) + P(1))$
$\approx 1 - 0.5578$
$= 0.442$ (using Poi(1.5)).

- Make sure you clarify which distribution you are using under these circumstances.

Q3 The mean annual income for all employees in a factory is £15 600. The standard deviation of income is £3100. Use an appropriate probability distribution to determine the probability that the total income of a random selection of 25 employees is more than £410 000.

- It would be useful to summarise the detail contained in this long introduction to the question.

$\Sigma X \sim N(25 \times 15\,600, 25 \times 3100^2)$.

- This demonstrates that you know about the sum of repeated observations.

Hence, $P(\Sigma X > 410\,000)$

$= 1 - \Phi\left(\frac{410\,000 - 390\,000}{5 \times 3100}\right)$

- Show the expression to be evaluated clearly.

$= 0.098$

- Don't overdo the accuracy.

Q4 In Basil's Spire Hotel, the probability of a guest being dissatisfied with the service is 0.45. A random sample of 40 guests were contacted a week after their stay and 10 expressed dissatisfaction.

State your hypotheses clearly, and use an appropriate approximation to test, at the five per cent level of significance, whether the number of dissatisfied customers is unusually low.

H_0: $p = 0.4$, H_1: $p < 0.4$

$B(40, 0.4) \approx N(16, 9.6)$ and the binomial $P(x \leqslant 10) \approx$ the normal $P(z < 10.5)$

$= 1 - \Phi\left(\frac{16 - 10.5}{\sqrt{9.6}}\right) = 0.0379$ which is significant at five per cent.

This suggests that the dissatisfaction rate is less than claimed.

How to score full marks

- You should realise that you are in a position to have gleaned the answer to this from the information provided up to this point.
- A one-sided test is indicated by 'low'.
- Make sure you are familiar with the conditions for various approximations to probability models.
- Clearly indicated use of the continuity correction.
- Clearly stated conclusion.

Questions to try

Q1 'Bless My Dear Aunt Sally' is a pneumonic that many students learn in primary school, but not many children actually have an aunt called Sally. A nationwide survey revealed that five per cent of children did. Find an approximate value for the probability that, in a random sample of 120 primary school pupils, exactly seven have an aunt called Sally.

Q2 In a college laboratory, radioactive events occur randomly at an average rate of 25 events per minute. Obtain an estimate of the probability that there will be 26 radiation events in the next minute.

Q3 Market research reveals that 38 per cent of readers enjoy the features section of a newspaper. Use an appropriate approximation to find the probability that at least 400 readers from a random sample of 1000 enjoyed the feature in today's paper.

Q4 The amount of money a typical shopper spends in a supermarket has a mean value of £10.52 and a standard deviation of £4.20. Obtain an estimate of the probability that the next 20 customers in the supermarket spend a total of at least £200.

Answers can be found on page 111.

19 Continuous random variables

Key points to remember

- **Density functions** and **distribution functions** feature prominently in the study of continuous random variables.
- The **probability density function**, $f(x)$, provides a model for the **relative frequency histogram**.

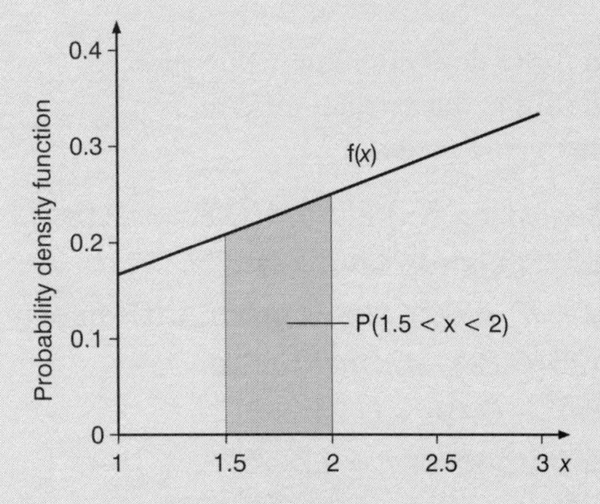

- The **area under the density function** gives the **probability**.
- The area may be calculated using integration, e.g.
  $P(1.5 < x < 2) = \int_{1.5}^{2} f(x)dx$
- The total area must be exactly equal to 1 (unity), e.g.
  $\int_{1}^{3} f(x)dx = 1$
- No part of the **density function** is allowed to be negative.
- The **cumulative distribution function**, $F(x)$, is the integral of the density function, e.g.
  $F(x) = \int_{1}^{x} f(t)dt$
- Probabilities may be determined using the **distribution function**, e.g.
  $P(1.5 < x < 2) = F(2) - F(1.5)$
- **Distribution tables** are provided for some common continuous random variables.

Formulae you must know

- $P(x < \text{median}) = 0.5$
- The mean of the distribution
  $\mu = E(x) = \int xf(x)dx$
- The variance of the distribution
  $\sigma^2 = E(x^2) - \mu^2 = \int x^2 f(x)dx - \mu^2$
- The standard deviation = $\sqrt{\text{variance}}$

Don't make these mistakes ...

Don't forget to differentiate the distribution function to obtain the density function.

Don't try to solve problems without sketching $f(x)$.

Don't assume that $\int xf(x)dx = xF(x)$.

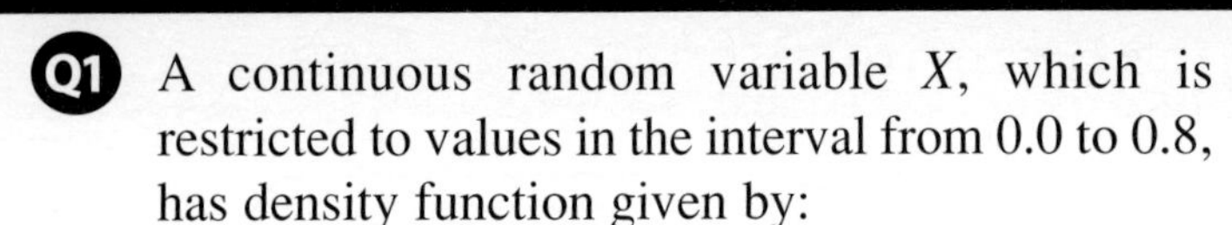

How to score full marks

Q1 A continuous random variable X, which is restricted to values in the interval from 0.0 to 0.8, has density function given by:

$f(x) = a(0.8 - x)$ for $0 < x < 0.8$

(a) Carefully determine the value of a.
(b) Determine the median of X.
(c) What is the mean of X?

(a) Integrating f(x) gives $a(0.8x - 0.5x^2)$, and this leads to $a = 3.125$.

(b) $P(x < \eta) = 0.5$, therefore η satisfies:

$3.125(0.8\eta - 0.5\eta^2) = 0.5$

$0.5\eta^2 - 0.8\eta + 0.16 = 0$

$\eta = 0.2343$ or $\eta = 1.3657$

So the median is 0.234.

- Note words such as 'carefully'; although this answer is right, the method does not show the processes involved.
- A sketch of the density function may have made the work easier.

(c) Integrating xf(x) over the interval 0 to 0.8 gives the mean as 0.2667.

- Don't make this difficult, especially if you can use your calculator functions.

Q2 The lifetime (x months) of a torch bulb is modelled by a probability density function which has cumulative distribution function given by:

$F(x) = ax^2(7.5 - x)$ for $0 < x < 5$.

- Notice that this is the distribution function.

(a) Show that $a = 0.016$.
(b) Determine the modal lifetime.
(c) A torch has just had a new bulb fitted. What is the probability that it will fail within three months?

(a) $F(5) = 1$, so $62.5a = 1$

Hence a is 0.016.

- Make sure you give plenty of indication of method when the answer is provided as in this case.

(b) $f(x) = 0.016(15x - 3x^2)$

The maximum is where

$f'(x) = 0.24 - 0.096x$

$= 0 \Rightarrow$ the mode is 2.5.

- Mode requires knowledge of the density function.
- A sketch may have been equally acceptable, and possibly easier!

(c) $F(3) = 0.648$

- Don't make this difficult – use the information given in the question.

Q3 The continuous random variable, T, has probability density function $f(t)$ given by:

$f(t) = 1.25t^{-2}$ for $1 < t < 5$
and $f(t) = 0$ elsewhere.

(a) Sketch $f(t)$.
(b) Determine the mean and the variance of T.
(c) Calculate the interquartile range of T.

(a)

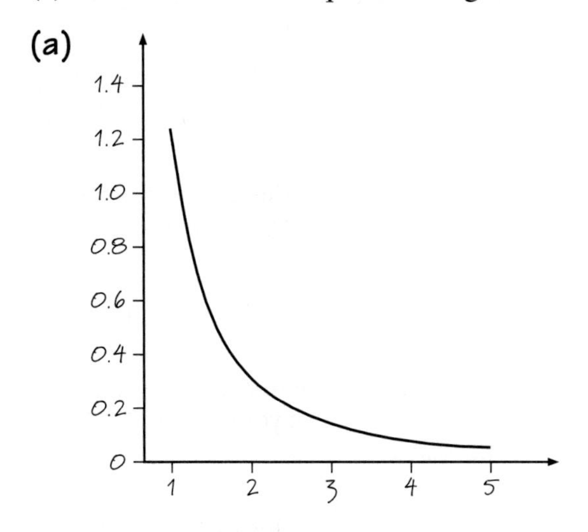

(b) Mean = $\int x \times 1.25x^{-2}dx = 1.25\ln 5 = 2.012$

Variance = $\int x^2 \times 1.25x^{-2}dx - (2.012)^2$

$= 0.9527$

(c) Lower quartile: $\int 1.25x^{-2}dx = 0.25$ gives

$q_1 = 1.25$.
Upper quartile comes from $\int 1.25x^{-2}dx$

$= 0.75$ which leads to $q_3 = 2.5$.
Hence the interquartile range is 1.25.

How to score full marks

- Don't be put off by the use of variables other than x.
- You need to be up to speed with your pure maths, especially integration giving a natural logarithm.
- Make full use of your calculator, especially its graphing facilities to help you with sketching.
- Don't be afraid of changing the variate in a definite integral if this helps.
- Don't forget the definition of variance used here: 'mean of squares minus square of mean'.
- Always refer back to the question to check that you have answered it fully.

Questions to try

Q1 A continuous random variable, X, has probability density function, f, given by:

$$f(x) = kx(4 - x)^2 \qquad \text{for } 0 \leqslant x \leqslant 4$$
$$\text{and} \quad f(x) = 0 \qquad \text{elsewhere.}$$

(a) Find the value of k and sketch a graph of $f(x)$.
(b) Determine the expected value of X.
(c) Calculate the modal value of X.

Q2 A model is proposed for the density function of the distribution of lengths, X m, of pieces of unused pipe in a wind-chime manufacturer's workshop:

$$f(x) = kx^{-2} \qquad \text{for } 0.1 \leqslant x \leqslant 1.0$$
$$\text{and} \quad f(x) = 0 \qquad \text{elsewhere.}$$

(a) Show carefully that $k = \frac{1}{9}$.
(b) Determine the distribution function for the lengths.
(c) Give an estimate of the probability that the length of a randomly chosen piece of pipe exceeds half a metre.
(d) Calculate the median and the quartile lengths.

Q3 A continuous random variable, X, has distribution function given by:

$$F(x) = \tfrac{1}{8}(x - a)(b - x) \qquad \text{when } 2 \leqslant x \leqslant 4$$
$$\text{and} \quad F(x) = 0 \qquad \text{elsewhere.}$$

(a) Justify the assertion that $a = 2$ and $b = 8$.
(b) Determine the probability density function for X.
(c) Calculate the median of X.
(d) One simple definition of natural skewness involves determining the sign of the expression 'mean – median'. Use this definition to decide whether this distribution is positively skewed, negatively skewed or is symmetrical.

Q4 A continuous random variable, T, has density function given by:

$$f(t) = \frac{24}{t^3} \qquad \text{when } 3 \leqslant t \leqslant 6$$
$$\text{and} \quad f(t) = 0 \qquad \text{elsewhere.}$$

(a) Determine the mean and the standard deviation of T.
(b) Obtain an expression for the distribution function $F(t)$.
(c) Determine the probability, $P(2 < t < 5)$.

Answers can be found on page 112.

Answers | **How to solve these questions**

1 The binomial theorem

Q1 $\binom{5}{4} \times 2^1 \times (3x)^4 = \frac{5 \times 4 \times 3 \times 2 \times 1}{4 \times 3 \times 2 \times 1 \times 1} \times 2 \times 81x^4$

$= 810x^4$

You don't need to produce the whole expansion, as the question only asks for the x^4 term.

The coefficient of x^4 is 810.

State the coefficient clearly.

Q2 $\left(x - \frac{1}{x}\right)^3 = x^3 + \binom{3}{1}x^2 \times \left(-\frac{1}{x}\right) + \binom{3}{2}x \times \left(-\frac{1}{x}\right)^2 + \left(-\frac{1}{x}\right)^3$

$= x^3 - 3x + \frac{3}{x} - \frac{1}{x^3}$

Use the formula for $(a + b)^n$ with $a = x$ and $b = -\frac{1}{x}$. Be careful with the – signs as you simplify the expressions.

Q3 $(1 + x)^5 = 1 + \binom{5}{1}x + \binom{5}{2}x^2 + \binom{5}{3}x^3 + \binom{5}{4}x^4 + x^5$

$= 1 + 5x + 10x^2 + 10x^3 + 5x^4 + x^5$

Use the formula for $(a + b)^n$ with $a = 1$ and $b = x$.

$(1 + y + y^2)^5 = 1 + 5(y + y^2) + 10(y + y^2)^2 + 10(y + y^2)^3 + 5(y + y^2)^4 + (y + y^2)^5$

Replace each x in the expansion by $y + y^2$.

$= 1 + 5y(1 + y) + 10y^2(1 + y)^2 + 10y^3(1 + y)^3 + 5y^4(1 + y)^4 + y^5(1 + y)^5$

$= \ldots + 10y^4 + \ldots + 10y^3 \times 3y + \ldots + 5y^4 \times 1 + \ldots$

$= \ldots + 45y^4 + \ldots$

Note that $y, y^2, y^3, \ldots$ can be taken as factors from each bracket. Only calculate the coefficients of the terms that will give y^4, but don't miss any out.

The coefficient of y^4 is 45.

Q4 $(1 - ax)^n = 1 + n(-ax) + \frac{n(n-1)}{2}(-ax)^2 + \ldots$

Generate the first three terms of the expansion, using

$(1 + ax)^n = 1 + nax + \frac{n(n-1)}{1 \times 2}(ax)^2 + \frac{n(n-1)(n-2)}{6}(ax)^3 + \ldots + \frac{n(n-1)(n-2)\ldots(n-r+1)}{r!}(ax)^r + \ldots$

$na = 20 \Rightarrow a = \frac{20}{n}$

Equate the coefficients of x in both expressions to give one equation relating a and n.

Repeat with coefficients of x^2 to form a second equation.

$\frac{n(n-1)}{2}a^2 = 160$

Combine the two equations to eliminate a and then solve for n.

$\frac{n(n-1)}{2}\left(\frac{20}{n}\right)^2 = 160$

$200(n - 1) = 160n$

$n = 5$

$a = \frac{20}{5} = 4$

Use the value of n to find a, by substituting it back into one of the earlier equations.

Q5

(a) $(p + qx)^3 = p^3 + \binom{3}{1}p^2qx + \binom{3}{2}p(qx)^2 + (qx)^3$

$= p^3 + 3p^2qx + 3pq^2x^2 + q^3x^3$

Use the standard formula to expand $(p + qx)^3$.

(b) $(1 + 2x)(p + qx)^3$

$= (1 + 2x)(p^3 + 3p^2qx + 3pq^2x^2 + q^3x^3)$

$= p^3 + (2p^3 + 3p^2q)x + (6p^2q + 3pq^2)x^2 + (q^3 + 6pq^2)x^3 + \ldots$

Multiply by $(1 + 2x)$, finding terms up to and including x^3.

$p^3 = 8 \Rightarrow p = 2$

Compare the two constant terms to find the value of p.

$30 = 6p^2q + 3pq^2$
$30 = 24q + 6p^2$
$0 = q^2 + 4q - 5$
$q = -5$ or $q = 1$

Compare the coefficients of x^2 to find the two possible values of q, given that $p = 2$.

$q = 1$ as $a > 0$

Select the value of q for which a is greater than zero.

$a = 2p^3 + 3p^2q = 16 + 12 = 28$

Compare the coefficients of x and use the values of p and q to find a.

$b = q^3 + 6pq^2 = 1 + 12 = 13$

Find the value of b in the same way.

Q6 $(1-2x)^{-\frac{1}{2}} = 1 + (-\frac{1}{2}) \times (-2x) + \frac{-\frac{1}{2}(-\frac{1}{2}-1)}{2} \times (-2x)^2 + \ldots$

$= 1 + x + \frac{3}{2}x^2 + \ldots$

Generate the first three terms of the expansion, using $(1+x)^n = 1 + n(x) + \frac{n(n-1)}{2}(x)^2 + \ldots$ substituting $(-2x)$ for x. Take care when you are simplifying.

The expansion is valid for $|2x| < 1$.
$-1 < 2x < 1 \Rightarrow -\frac{1}{2} < x < \frac{1}{2}$

Recall that in the formula book it states that the expansion of $(1+x)^n$ is true for $|x| < 1$, in this case replace $|x|$ by $|2x|$.

Q7

(a) $(1-3x)^{\frac{1}{3}} = 1 + \frac{1}{3} \times (-3x) + \frac{\frac{1}{3}(\frac{1}{3}-1)}{2} \times (-3x)^2 + \frac{\frac{1}{3}(\frac{1}{3}-1)(\frac{1}{3}-2)}{6} \times (-3x)^3$

$= 1 - x - x^2 - \frac{5}{3}x^3 - \ldots$

Generate the first three terms of the expansion, using $(1+x)^n = 1 + nx + \frac{n(n-1)}{2}x^2 + \ldots$ substituting $(-3x)$ for x.

Take care simplifying.

(b) $(1 - 3 \times 10^{-3})^{\frac{1}{3}} = 0.997^{\frac{1}{3}}$

$= 1 - 0.001 - 0.000\,001 - 0.000\,000\,001\,67$
$= 0.998\,998\,998$

Substitute $x = 10^{-3}$ to obtain the cube root of 0.997.

$\sqrt[3]{997} = \sqrt[3]{1000 \times 0.997} = 10 \times 0.998\,998\,998$
$= 9.989\,989\,98$

Multiply by the cube root of 1000 (10), to find the cube root of 997. Check the answer on your calculator.

Q8 $\frac{1}{1-x} + \frac{1}{1-2x}$

$= (1-x)^{-1} + (1-2x)^{-1}$
$= 1 + x + x^2 + x^3 + \ldots + 1 + 2x + 4x^2 + 8x^3 + \ldots$
$= 2 + 3x + 5x^2 + 9x^3 + \ldots$

Rewrite each fraction as a bracket raised to a power, then expand each bracket in the normal way and add the results.

$|2x| < 1 \Rightarrow -1 < 2x < 1 \Rightarrow -\frac{1}{2} < x < \frac{1}{2}$

Remember that one expansion is valid for $|x| < 1$ and the other is valid for $|2x| < 1$.

Q9

(a) $\sqrt{1+x^2} = (1+x^2)^{\frac{1}{2}} = 1 + \frac{1}{2} \times x^2 + \frac{\frac{1}{2}(\frac{1}{2}-1)}{2} \times (x^2)^2 + \ldots$

$= 1 + \frac{1}{2}x^2 - \frac{1}{8}x^4 + \ldots$

Use the expansion formula, but replace x by x^2.

(b) $\int_0^{0.4} \sqrt{1+x^2}\,dx \approx \int_0^{0.4} (1 + \frac{1}{2}x^2 - \frac{1}{8}x^4)\,dx$

$= \left[x + \frac{1}{6}x^3 - \frac{1}{40}x^5\right]_0^{0.4}$

$= 0.410$ to 3 s.f.

Integrate the expression from part (a), to find an approximate value for the integral.

Give the answer to 3 significant figures, as specified in the question.

2 Trigonometry

Q1 $\sin(x + 30°) \equiv \sin x\cos 30° + \cos x\sin 30°$

$\equiv \frac{\sqrt{3}}{2}\sin x + \frac{1}{2}\cos x$

Use the identities for $\sin(A+B)$ and $\cos(A+B)$ to expand $\sin(x+30°)$ and $\cos(x+30°)$.

$\cos(x + 30°) \equiv \cos x\cos 30° - \sin x\sin 30°$

$\equiv \frac{\sqrt{3}}{2}\cos x - \frac{1}{2}\sin x$

$\sin(x + 30°) + \sqrt{3}\cos(x + 30°)$

$\equiv \frac{\sqrt{3}}{2}\sin x + \frac{1}{2}\cos x + \frac{3}{2}\cos x - \frac{\sqrt{3}}{2}\sin x \equiv 2\cos x$

When $x = 15°$:

Substitute $x = 15°$.

$2\cos 15° = \sin 45° + \sqrt{3}\cos 45°$

$= \frac{\sqrt{2}}{2} + \sqrt{3} \times \frac{\sqrt{2}}{2} = \frac{\sqrt{2}}{2}(1 + \sqrt{3})$

Do not use your calculator to simplify because the question needs your answer to be in surd form.

Q2

(a) $\tan x + \cot x \equiv \frac{\sin x}{\cos x} + \frac{\cos x}{\sin x} \equiv \frac{\sin^2 x + \cos^2 x}{\sin x\cos x}$

$\equiv \frac{2}{\frac{1}{2}\sin 2x} \equiv \frac{2}{\sin 2x}$

Add the fractions and use the two identities $\sin^2 x + \cos^2 x = 1$ and $\sin 2x = 2\sin x\cos x$ to simplfy your answer.

(b) $\tan x + \cot x > 4$

$\frac{2}{\sin 2x} > 4$

$\sin 2x < \frac{1}{2}$

$0 < x < \frac{\pi}{12}$

You can multiply by $\sin 2x$ because $\sin 2x > 0$ for $0 < x < \frac{\pi}{4}$.

(c) Volume $= \pi\int_{\frac{\pi}{8}}^{\frac{\pi}{6}} y^2 dx = \pi\int_{\frac{\pi}{8}}^{\frac{\pi}{6}} \frac{4}{\sin^2 2x}dx$

$= 4\pi\int_{\frac{\pi}{8}}^{\frac{\pi}{6}} \text{cosec}^2 2x\,dx = 4\pi\left[-\frac{1}{2}\cot 2x\right]_{\frac{\pi}{8}}^{\frac{\pi}{6}}$

$= 2\pi(-\cot\frac{\pi}{3} + \cot\frac{\pi}{4}) = \frac{2\pi}{3}(3 - \sqrt{3})$

Replace y by $\tan x$ and $\cot x$ by $\frac{2}{\sin 2x}$. Use the fact that $\frac{1}{\sin 2x} = \text{cosec}2x$ and $\int \text{cosec}^2 u\,du = \cot u$.

Remember $\cot\frac{\pi}{4} = \frac{1}{\tan\frac{\pi}{4}} = 1$ and $\cot\frac{\pi}{3} = \frac{1}{\tan\frac{\pi}{3}} = \frac{1}{\sqrt{3}}$.

Give your answer in surd form to obtain full marks.

Q3

(a) $\sin 4x \equiv \sin(2 \times 2x) \equiv 2\sin 2x\cos 2x$
$\equiv 4\sin x\cos x\cos 2x$

Use the identity $\sin 2\theta \equiv 2\sin\theta\cos\theta$ twice.

(b) $\sin x = \frac{3}{5}$ $\quad \cos x = \frac{4}{5}$

$\cos 2x = 1 - 2\sin^2 x = 1 - 2(\frac{9}{25}) = \frac{7}{25}$

$\sin 4x = 4 \times \frac{3}{5} \times \frac{4}{5} \times \frac{7}{25} = \frac{336}{625}$

Recognise the 3, 4, 5 triangle and use the identity $\cos 2x = 1 - 2\sin^2 x$.

Q4

(a) $R\cos(x - A) \equiv R\cos x\cos A + R\sin x\sin A$

Compare this with $\cos x$ and $\sin x$ to give:

$\left.\begin{array}{l} R\cos A = 1 \\ R\sin A = 1 \end{array}\right\} R = \sqrt{2}$

and $\tan A = 1 \Rightarrow A = \frac{\pi}{4}$

Use the compound angle formula to expand $R\cos(x - A)$. Solve these two equations to find the values of R and A.

(b) $\cos x + \sin x = \sqrt{2}\cos(x - \frac{\pi}{4})$
$= 1$

$\cos(x - \frac{\pi}{4}) = \frac{1}{\sqrt{2}}$

$x - \frac{\pi}{4} = \pm\frac{\pi}{4} + 2n\pi$

$x = 2n\pi$ or $x = \frac{\pi}{2} + 2n\pi$

Use the result that if $\cos\theta = \frac{1}{\sqrt{2}}$ then $\theta = \frac{\pi}{4} + 2n\pi$ and $\theta = -\frac{\pi}{4} + 2n\pi$.

(c) The greatest value of $\cos x$ and $\sin x$ is the greatest value of $\sqrt{2}\cos(x - \frac{\pi}{4}) = \sqrt{2}$.

The greatest value of $\cos\theta$ is 1.

Q5 $\sin(\theta + 2\theta) = \sin\theta\cos 2\theta + \cos\theta\sin 2\theta$
$= \sin\theta(1 - 2\sin^2\theta) + \cos\theta(2\sin\theta\cos\theta)$
$= \sin\theta - 2\sin^3\theta + 2\sin\theta\cos^2\theta$
$= \sin\theta - 2\sin^3\theta + 2\sin\theta(1 - \sin^2\theta)$
$= 3\sin\theta - 4\sin^3\theta$

Use the trigonometric identity for $\sin(A + B)$. Remember that $\cos 2\theta = 1 - 2\sin^2\theta$, $\sin 2\theta = 2\sin\theta\cos\theta$ and $\cos^2\theta = 1 - \sin^2\theta$.

$\sin 3\theta = 2\sin\theta$
$3\sin\theta - 4\sin^3\theta = 2\sin\theta$
$\sin\theta - 4\sin^3\theta = 0$
$\sin\theta(1 - 4\sin^2\theta) = 0$

Be careful not to divide by $\sin\theta$ since $\sin\theta = 0$ is a possible solution.

$\sin\theta = 0$ or $(1 - 4\sin^2\theta) = 0$
$\Rightarrow \sin\theta = 0$ or $\sin\theta = \frac{1}{2}$ or $\sin\theta = -\frac{1}{2}$

$\theta = 0, 180°, 360°$ or $\theta = 30°, 150°$ or $\theta = 210°, 330°$.
$\theta = 0, 30°, 150°, 180°, 210°, 330°, 360°$

Q6

(a) $\sin 2x + 2\cos^2 x = 0$
$\sin 2x + \cos 2x + 1 = 0$

Use the identity $\cos 2x \equiv 2\cos^2 x - 1$ to write $f(x)$ is terms of $\sin 2x$ and $\cos 2x$.

$R\sin(2x + A) + 1 = 0$

Then $\left.\begin{array}{l} R\cos A = 1 \\ R\sin A = 1 \end{array}\right\} R = \sqrt{2}$

and $\tan A = 1 \Rightarrow A = \frac{\pi}{4}$

Express the trigonometric terms as a single sine function. Alternatively, use $\sin 2x \equiv 2\sin x\cos x$ and obtain $\cos x(\sin x + \cos x) = 0$, which you can then solve. This gives the same results but does not help with part (b).

$\sqrt{2}\sin(2x + \frac{\pi}{4}) = -1$

$2x + \frac{\pi}{4} = \frac{5\pi}{4} + 2n\pi$ or $2x + \frac{\pi}{4} = \frac{7\pi}{4} + 2n\pi$

$x = \frac{\pi}{2} + n\pi$ or $x = \frac{3\pi}{4} + n\pi$
where $n = -3, -2, -1, 0, 1, 2, 3, \ldots$

Write down the general solution.

Answers / How to solve these questions

(b) $f(x) = \sqrt{2}\sin(2x + \frac{\pi}{4}) + 1$

$1 - \sqrt{2} \leqslant f(x) \leqslant 1 + \sqrt{2}$

Use the fact that $\sin\theta$ varies between –1 and +1.

(c) Area $= \int_{-\frac{\pi}{4}}^{\frac{\pi}{2}} f(x)dx$

$= \int_{-\frac{\pi}{4}}^{\frac{\pi}{2}} (1 + \sqrt{2}\sin(2x + \frac{\pi}{4}))dx = \left[x - \frac{\sqrt{2}}{2}\cos(2x + \frac{\pi}{4})\right]_{-\frac{\pi}{4}}^{\frac{\pi}{2}}$

$= \left(\frac{\pi}{2} - \frac{\sqrt{2}}{2} \times \cos(\pi + \frac{\pi}{4})\right) - \left(-\frac{\pi}{4} - \frac{\sqrt{2}}{2}\cos(-\frac{\pi}{2} + \frac{\pi}{4})\right)$

$= \frac{3\pi}{4} - \frac{\sqrt{2}}{2} \times \left(-\frac{1}{\sqrt{2}}\right) + \frac{\sqrt{2}}{2} \times \frac{1}{\sqrt{2}} = \frac{3\pi}{4} + 1$

To find the limits of integration, find the smallest possible value of x for which $f(x) = 0$ (point B) and the largest negative value of x for which $f(x) = 0$ (point A). The value $n = 0$ gives $x = \frac{\pi}{2}$ (point B) and $n = -1$ gives $x = -\frac{\pi}{4}$ (point B).

Q7

(a) QR = QB + BR = ABcos∠ABQ + BCcos$x$

Since ∠ABC is 90°, ∠ABQ = 90° – x.

cos∠ABQ = cos(90° – x) = sin$x$

AB = 18 and BC = 27

QR = 18sin$x$ + 27cos$x$ = 30

9cos$x$ + 6sin$x$ = 10

(b) $R\cos(x - \alpha) = R\cos x\cos\alpha + R\sin x\sin\alpha$

$R\cos\alpha = 9$ and $R\sin\alpha = 6$

$R^2 = 9^2 + 6^2 = 117$ and $\tan\alpha = \frac{6}{9}$

$R = 10.82$ and $\alpha = 33.69°$

Use the identity for cos(A – B) and compare your result with the formula from part (a). Solve for R and α. Give the answer correct to 2 decimal places as stated in the question.

(c) $10.82\cos(x - 33.69°) = 10$

$\cos(x - 33.69°) = 10 \div 10.82$

$x - 33.69° = 22.45°$

$x = 56.1°$

Give the answer correct to the nearest tenth of a degree.

3 Differentiation techniques

Q1

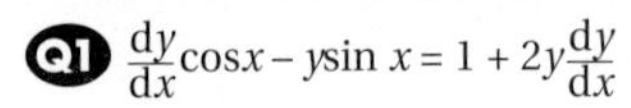

$\frac{dy}{dx}\cos x - y\sin x = 1 + 2y\frac{dy}{dx}$

Use implicit differentiation with the product rule applied to $y\cos x$.

$\frac{dy}{dx} = \frac{1 + y\sin x}{\cos x - 2y}$

Make $\frac{dy}{dx}$ the subject of the expression.

At the point where $x = 0$ and $y = 1$:

$\frac{dy}{dx} = \frac{1 + 0}{\cos 0 - 2} = -1$

Substitute the values of x and y and simplify.

Q2

$\frac{dV}{dx} = \frac{(x + 9) \times 15x^{-\frac{1}{2}} - 30\sqrt{x}}{(x + 9)^2}$

$= \frac{135 - 15x}{\sqrt{x}(x + 9)^2}$

First differentiate using the quotient rule. Multiply the numerator and denominator of the fraction by $\sqrt{x}$ so that you can simplify.

$0 = 135 - 15h$

$h = 9$ cm

$V = \frac{30\sqrt{9}}{18} = 5\text{ cm}^3$

Replace x by h. For $\frac{dV}{dx} = 0$, the numerator of the fraction must be zero. Solve the equation to find h and then substitute $x = 9$ into the formula for V.

Q3

$\frac{dy}{dx} = \frac{(6 - x) \times 2 - (2x + 3) \times (-1)}{(6 - x)^2}$

$= \frac{15}{(6 - x)^2}$

Differentiate, using the quotient rule, then simplify the result.

At $x = 5$: $\frac{dy}{dx} = 15$

Find the gradient of the curve when $x = 5$.

$y = -\frac{x}{15} + c$

For the point where $x = 5$, $y = 13$:

$13 = -\frac{1}{3} + c \Rightarrow c = \frac{40}{3}$

$y = -\frac{x}{15} + \frac{40}{3}$

The gradient of the normal will be $-\frac{1}{15}$, so write down the equation of the normal using $y = mx + c$ and find the value of c.

Answers

Q4 $\frac{dy}{dx} = (x-4)(-e^{-x}) + 1 \times e^{-x} = (5-x)e^{-x}$

$0 = (5-x)e^{-x}$

$x = 5 \Rightarrow y = e^{-5}$

$\frac{d^2y}{dx^2} = (x-6)e^{-x}$

$\frac{d^2y}{dx^2} < 0$ when $x = 5$

$\therefore$ the point is a local maximum.

Q5 $\frac{dx}{dt} = 4t,\ \frac{dy}{dt} = 3t^2$

$\frac{dy}{dx} = \frac{3t^2}{4t} = \frac{3t}{4}$

When $t = 2$: $\frac{dy}{dx} = \frac{3 \times 2}{4} = \frac{3}{2}$

Q6

(a) $\frac{dx}{dt} = 12\sin^2 t\cos t$

$\frac{dy}{dt} = -2\sin 2t = -4\sin t\cos t$

$\frac{dx}{dy} = \frac{12\sin^2 t\cos t}{-4\sin t\cos t} = -3\sin t$

(b) Gradient $= 3\sin\frac{\pi}{6} = \frac{3}{2}$

$y = \frac{3}{2}x + c$

$t = \frac{\pi}{6} \Rightarrow x = \frac{1}{2},\ y = \frac{1}{2}$

$\frac{1}{2} = \frac{3}{2} \times \frac{1}{2} + c \Rightarrow c = -\frac{1}{4}$

$y = \frac{3}{2}x - \frac{1}{4}$

Q7 $\frac{dx}{d\theta} = a\cos\theta$

$\frac{dy}{d\theta} = a\cos\theta - a\theta\sin\theta$

$\frac{dy}{dx} = \frac{a\cos\theta - a\theta\sin\theta}{a\cos\theta} = \frac{\cos\theta - \theta\sin\theta}{\cos\theta}$

$0 = \frac{\cos\theta - \theta\sin\theta}{\cos\theta}$

$\cos\theta - \theta\sin\theta = 0$

$\frac{\sin\theta}{\cos\theta} = \frac{1}{\theta}$

$\tan\theta = \frac{1}{\theta}$

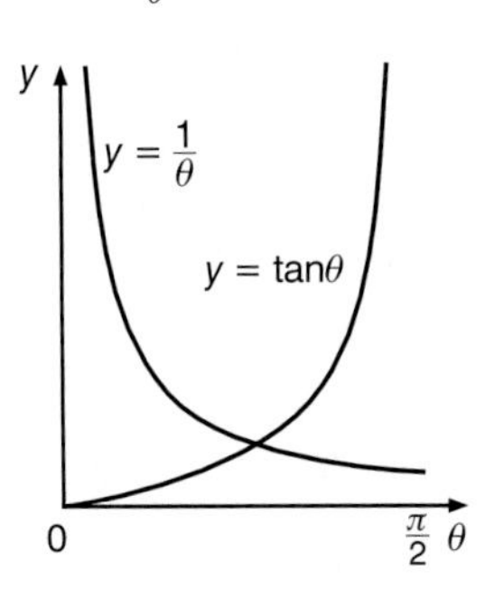

$\theta = \frac{\pi}{4} \Rightarrow \tan\theta = 1$ and $\frac{1}{\theta} = \frac{4}{\pi} > 1$.

At this point $\frac{1}{\theta} > \tan\theta$.

So the solution is greater than $\frac{\pi}{4}$ because $\tan\theta$ is increasing and $\frac{1}{\theta}$ is decreasing.

How to solve these questions

Q4 Differentiate, using the product rule, then simplify the result.

Find the value of x when $\frac{dy}{dx} = 0$, then find the value of y.

Use $\frac{d^2y}{dx^2}$ (the second derivative) to determine the nature of the stationary point.

As this is negative, the stationary point is a local maximum.

Q5 Find $\frac{dx}{dt}$ and $\frac{dy}{dt}$, then use the formula $\frac{dy}{dx} = \frac{dy}{dt} / \frac{dx}{dt}$.

Then substitute $t = 2$ to find the required gradient.

Q6 (a) Find $\frac{dx}{dt}$ and $\frac{dy}{dt}$, then use the formula $\frac{dy}{dx} = \frac{dy}{dt} / \frac{dx}{dt}$.

You need to use the identity $\sin 2x = 2\sin x\cos x$ to reach the required result.

(b) You can find the gradient of the normal, using the facts that the product of the two gradients (tangent and normal) is –1, and $\frac{1}{\frac{dy}{dx}} = \frac{dx}{dy}$.

You need to find the values of x and y when $t = \frac{\pi}{6}$.

Then use this information to find an equation for the normal in the form $y = mx + c$.

Q7 You should use parametric differentiation for this.

Use the product rule for $\frac{dy}{d\theta}$.

Cancelling the as, while not vital, keeps the work neat.

The gradient of the curve will be zero when $\frac{dy}{dx} = 0$.

Rearrange the equation, using $\tan\theta = \frac{\sin\theta}{\cos\theta}$.

Draw the graphs, showing $\frac{1}{\theta}$ very large close to $\theta = 0$ and decreasing over the interval, and $\tan\theta$ increasing from 0 and becoming very large as θ approaches $\frac{\pi}{2}$.

You need to present an argument based on a comparison of the values when $\theta = \frac{\pi}{4}$. The comparison shows that this value of θ is to the left of the intersection because $\frac{1}{\theta} > \tan\theta$.

Answers / How to solve these questions

Q8 $\frac{dy}{dx} = \frac{\sqrt{x^2+1} \times 1 - x \times x(x^2+1)^{-\frac{1}{2}}}{x^2+1}$

Differentiate, using the quotient rule.

$= \frac{x^2+1-x^2}{(x^2+1)^{\frac{3}{2}}} = \frac{1}{(x^2+1)^{\frac{3}{2}}}$

You can simplify, after multiplying the numerator and denominator of the fraction by $\sqrt{x^2+1}$.

$\int \frac{2}{(x^2+1)^{\frac{3}{2}}} dx = \frac{2x}{\sqrt{x^2+1}} + c$

You should spot the similarity between the derivative you have just obtained and the function to be integrated. Don't forget to include the factor of 2 and the constant of integration.

4 Integration

Q1 $V = \pi \int_0^1 (x - x^2)^2 dx = \pi \int_0^1 (x^4 - 2x^3 + x^2) dx$

Use the formula $V = \pi \int y^2 dx$. The limits of integration are 0 and 1, which are the solutions of the equation $x - x^2 = 0$.

$= \pi \left[\frac{x^5}{5} - \frac{x^4}{2} + \frac{x^3}{3}\right]_0^1$

Be careful not to miss out any terms when expanding the brackets.

$d = \frac{\pi}{30}$

Give your answer in terms of π.

Q2

(a) $A = \int_1^5 \frac{2}{x+1} dx = \left[2\ln(x+1)\right]_1^5$

Note that $\int \frac{1}{x+1} dx = \ln(x+1)$.

$= 2\ln6 - 2\ln2 = \ln9$

Use the laws of logarithms to simplify the answer. Don't show any decimal values in your working.

(b) $V = \pi \int_1^5 \frac{4}{(x+1)^2} dx$

Use the formula $V = \pi \int y^2 dx$.

$= \pi \left[\frac{-4}{x+1}\right]_1^5$

Note that $\int \frac{1}{(x+1)^2} dx = \frac{-1}{x+1}$.

$= \frac{4\pi}{3}$

Simplify your answer without using a calculator.

Q3 $u = \sin x$

Use the substitution $u = \sin x$.

$\frac{du}{dx} = \cos x$ and $\frac{dx}{du} = \frac{1}{\cos x}$

First differentiate u to find a relationship between du and dx.

$\int \sin^3 x \sin 2x dx = \int 2\sin^4 x \cos x dx$

Use the formula $\sin 2x = 2\sin x \cos x$ to simplify the integral.

$= \int 2u^4 \cos x \frac{1}{\cos x} du = \int 2u^4 du$

Replace dx by $\frac{dx}{du} du$.

$= \frac{2u^5}{5} + c$

Integrate with respect to u.

$= \frac{2}{5}\sin^5 x + c$

Change the answer back, so that it is in terms of x.

Q4 $u = \ln 4x \Rightarrow \frac{du}{dx} = \frac{1}{x}$

$\frac{dv}{dx} = x^2 \Rightarrow v = \frac{1}{3}x^3$

You need to use integration by parts. In this case, differentiate $\ln 4x$ and integrate x^2. This produces an integral that only contains powers of x.

$\int_1^2 x^2 \ln 4x dx = \left[\frac{x^3}{3}\ln 4x\right]_1^2 - \int_1^2 \frac{x^2}{3} dx$

Use the laws of logarithms to obtain the required answer.

$= \frac{8}{3}\ln 8 - \frac{1}{3}\ln 4 - \left[\frac{x^3}{9}\right]_1^2 = \frac{22}{3}\ln 2 - \frac{7}{9}$

$= \frac{66\ln 2 - 7}{9}$

Q5 $u = x^3$

First you need to find a suitable substitution. In this case $u = x^3$ is a good choice because it will just produce $\cos u$ which can easily be integrated.

$\frac{du}{dx} = 3x^2 \Rightarrow \frac{dx}{du} = \frac{1}{3x^2}$

$\int x^2 \cos x^3 dx = \int \frac{\cos u}{3} du = \frac{\sin u}{3} + c$

Differentiate u and then substitute, also replacing dx by $\frac{dx}{du} du$.

$= \frac{\sin x^3}{3} + c$

When you have finished the integration, express the answer in terms of x rather than u.

Answers	How to solve these questions

Q6 $u = \ln x \quad \Rightarrow \quad \frac{du}{dx} = \frac{1}{x}$

$\frac{dv}{dx} = x \quad \Rightarrow \quad v = \frac{1}{2}x^2$

$$\int_1^a x\ln x\,dx = \left[\frac{x^2}{2}\ln x\right]_1^a - \int_1^a \frac{1}{x} \times \frac{x^2}{2}dx$$
$$= \frac{a^2}{2}\ln a - \int_1^a \frac{x}{2}dx = \frac{a^2}{2}\ln a - \left[\frac{x^2}{4}\right]_1^a$$
$$= \frac{a^2}{2}\ln a + \frac{1-a^2}{4}$$

You will need to use integration by parts, in a similar way to question 4.

Q7

(a) **(i)** $\cos(A+B) + \cos(A-B) = \cos A\cos B - \sin A\sin B + \cos A\cos B + \sin A\sin B = 2\cos A\cos B$

(ii) $\cos 2A = \cos^2 A - \sin^2 A = 2\cos^2 A - 1$

$\cos^2 A = \frac{1}{2}(1 + \cos 2A)$

For the first identity simply add the expressions for $\cos(A + B)$ and $\cos(A - B)$. Simplifying gives the required result.

For the second identity use $\cos 2A = \cos(A + A)$ and simplify the result. You will also need to use $\sin^2 A + \cos^2 A = 1$.

(b) $$\int \cos 3x\cos x\,dx = \int \frac{1}{2}(\cos 4x + \cos 2x)dx = \frac{1}{8}\sin 4x + \frac{1}{4}\sin 2x + c$$

You need to use the result in part (a) with $A = 3x$ and $B = x$. This produces an expression that you can integrate.

(c) $x = \cos t$

$\frac{dx}{dt} = -\sin t$

Begin by using the substitution and then differentiating.

$$\int \frac{x^2}{(1-x^2)^{\frac{1}{2}}}dx = \int \frac{\cos^2 t}{(1-\cos^2 t)^{\frac{1}{2}}} \times (-\sin t)dt$$
$$= \int -\cos^2 t\,dt$$
$$= \int -\frac{1}{2}(1 + \cos 2t)dt$$
$$= -\frac{t}{2} - \frac{1}{4}\sin 2t + c$$

Substitute in the integral for t and replace dx by $\frac{dx}{dt}dt$. Note that the denominator simplifies to $\sin t$.

$$\int_0^{\frac{1}{2}} \frac{x^2}{(1-x^2)^{\frac{1}{2}}}dx = \left[-\frac{t}{2} - \frac{1}{4}\sin 2t\right]_{\frac{\pi}{2}}^{\frac{\pi}{3}} = \frac{\pi}{12} - \frac{\sqrt{3}}{8}$$

Once you have completed the integration work back from $x = \cos t$ to change the limits of integration, then substitute these values to obtain the final result.

5 Vectors

Q1

(a) $\vec{AB} = \mathbf{b} - \mathbf{a} = -6\mathbf{i} + 24\mathbf{j} + 12\mathbf{k}$

$(2\mathbf{i} - 3\mathbf{j} + 7\mathbf{k}).\mathbf{AB} = -12 - 72 + 84 = 0$

$\therefore$ the vectors are perpendicular.

Show that the scalar product is zero, then the vectors must be perpendicular.

(b) $\mathbf{b} - \mathbf{a} = -6\mathbf{i} + 24\mathbf{j} + 12\mathbf{k}$

$\mathbf{r} = 4\mathbf{i} + 2\mathbf{j} - \mathbf{k} + t(-6\mathbf{i} + 24\mathbf{j} + 12\mathbf{k})$

$= (4 - 6t)\mathbf{i} + (2 + 24t)\mathbf{j} - (12t - 1)\mathbf{k}$

Use $\mathbf{r} = \mathbf{a} + t(\mathbf{b} - \mathbf{a})$

Q2 $\vec{AB} = = \begin{pmatrix}1\\2\\-7\end{pmatrix}$

The vectors are given as column vectors, so $\begin{pmatrix}3\\2\\4\end{pmatrix} = 3\mathbf{i} + 2\mathbf{j} + 4\mathbf{k}$.

The direction of l_1 is $\begin{pmatrix}5\\1\\1\end{pmatrix}$.

$\begin{pmatrix}1\\2\\-7\end{pmatrix} \cdot \begin{pmatrix}5\\1\\1\end{pmatrix} = 5 + 2 - 7 = 0$

AB is perpendicular to l_1.

For intersection:

$\begin{pmatrix}3+5t\\2+t\\4+t\end{pmatrix} = \begin{pmatrix}4+2s\\4+s\\-3-2s\end{pmatrix}$

$\left.\begin{array}{l}3 + 5t = 4 + 2s\\2 + t = 4 + s\end{array}\right\} \Rightarrow \begin{array}{l}t = -1\\s = -3\end{array}$

Check: $4 + t = 3$ and $-3 - 2s = 3$

When $t = -1$: $\begin{pmatrix}3+5t\\2+t\\4+t\end{pmatrix} = \begin{pmatrix}-2\\1\\3\end{pmatrix}$

The point of intersection is $(-2, 1, 3)$.

Remember to use the third unused equation to check that the values of s and t are correct.

(a) $\overrightarrow{OA}.\overrightarrow{AB} = (3\mathbf{i} - \mathbf{j} + 2\mathbf{k}).(-4\mathbf{i} + 2\mathbf{j} + 7\mathbf{k})$
$= -12 - 2 + 14 = 0$
OA is perpendicular to AB.

(b) L_1: $\mathbf{r} = 3\mathbf{i} - \mathbf{j} + 2\mathbf{k} + s(-4\mathbf{i} + 2\mathbf{j} + 7\mathbf{k})$

(c) The lines intersect if:
$3\mathbf{i} - \mathbf{j} + 2\mathbf{k} + s(-4\mathbf{i} + 2\mathbf{j} + 7\mathbf{k}) = 8\mathbf{i} + \mathbf{j} - 6\mathbf{k} + \mu(\mathbf{i} - 2\mathbf{j} - 2\mathbf{k})$

$$\left.\begin{aligned} 3 - 4s &= 8 + \mu \\ -1 + 2s &= 1 - 2\mu \end{aligned}\right\} \Rightarrow \begin{aligned} \mu &= 3 \\ s &= -2 \end{aligned}$$
$2 + 7s = -6 - 2\mu$
Check: $2 + 7s = -12$
$-6 - 2\mu = -12$
Hence the position vector of the point of intersection is $11\mathbf{i} - 5\mathbf{j} - 12\mathbf{k}$.

(d) $\cos\theta = \dfrac{(-4\mathbf{i} + 2\mathbf{j} + 7\mathbf{k}).(\mathbf{i} - 2\mathbf{j} - 2\mathbf{k})}{\sqrt{69}\sqrt{9}} = \dfrac{-22}{3\sqrt{69}}$

$\theta = 28.0°$

Since $\cos\theta < 0$ the angle $\theta > 90°$. The acute angle is $180° - \theta$.

(a) $\begin{pmatrix}2\\1\\3\end{pmatrix} + s\begin{pmatrix}1\\3\\-5\end{pmatrix} = \begin{pmatrix}-4\\3\\5\end{pmatrix} + t\begin{pmatrix}1\\-2\\2\end{pmatrix}$

$$\left.\begin{aligned} 2 + s &= -4 + t \\ 1 + 3s &= 3 - 2t \end{aligned}\right\} \Rightarrow \begin{aligned} s &= -2 \\ t &= 4 \end{aligned}$$
$3 - 5s = 5 + 2t$ Check: $3 - 5s = 13$
$5 + 2t = 13$

The point of intersection is (0, –5, 13).

(b) $\begin{pmatrix}1\\3\\5\end{pmatrix}\cdot\begin{pmatrix}1\\-2\\2\end{pmatrix} = 5$

$\cos\theta = \dfrac{5}{\sqrt{35}\sqrt{9}} = \dfrac{5}{3\sqrt{35}}$

$\theta = 73.6°$

6 Proof

(a) $n = 2 \quad \Rightarrow \quad n^2 - n - 2 = 0$

Since this is a quadratic equation with two solutions, $n = 2$ and $n = -1$ you cannot write $\Leftrightarrow$.

(b) $ab = 0 \quad \Leftrightarrow \quad a = 0$ or $b = 0$

If $ab = 0$ then either $a = 0$ or $b = 0$, and if $a = 0$ or $b = 0$ then $ab = 0$.

(c) n is a multiple of 2 $\Leftarrow$ n is a multiple of 4

If $n = 6$ then the right-hand statement is false so you can only use $\Leftarrow$.

(d) $\sin\theta = \dfrac{1}{2} \quad \Leftarrow \quad \theta = \dfrac{\pi}{6}$

If $\sin\theta = \dfrac{1}{2}$, then $\theta = 2n\pi + \dfrac{\pi}{6}$ or $\theta = 2n\pi + \dfrac{5\pi}{6}$.

(e) m is odd $\Leftrightarrow$ $m = 2n + 1$ for some integer n

All the odd numbers can be written in the form $2n + 1$.

(a) If 4 divides into n^2 then 4 divides into n.
If $n^2 = 36$ then $n = 6$ but 4 does not divide into 6.

The converse of the statement is not true.

You just need to find one counter-example to show that the converse of the statement is not true.

(b) If $(x - a)$ is a factor of $P(x)$ then $P(a) = 0$.
If $(x - a)$ is a factor of $P(x)$ then
$P(x) = (x - a)Q(x)$ for some $Q(x)$
$P(a) = (a - a)Q(a) = 0$

The converse of the statement is true.

Give a direct proof of this result.

(c) If $\dfrac{dy}{dx} = \cos x$ then $y = \sin x$.

If $\dfrac{dy}{dx} = \cos x$ then $y = \sin x + c$
where c is some constant.

The converse of the statement is false.

Answers | **How to solve these questions**

(d) Every number of the form $6n \pm 1$ is a prime number greater than 3.

If $n = 4$ then $6n + 1 = 25$ which is not prime.

The converse of the statement is false.

You just need to find one counter-example to show that the converse of the statement is not true.

Q3 If p is a prime number greater than 3 then p can be written in the form $6n \pm 1$ for some integer n.

In any proof, you have to start somewhere! The starting point is usually given in an examination question. In practice you might need to find the appropriate starting point yourself.

$$p = 6n + 1 \Rightarrow p^2 - 1 = (6n + 1)^2 - 1 = 36n^2 + 12n + 1 - 1$$
$$= 36n^2 + 12n = 12n(3n + 1)$$

Now use the method of direct proof for $p = 6n + 1$ and $p = 6n - 1$ separately.

If n is even then $n = 2m$ and $p^2 - 1 = 24m(6m + 1)$ and is divisible by 24.

For $12n(3n + 1)$ you need to show that $n(3n + 1)$ is even. Consider the cases for n even and n odd.

If n is odd then $n = 2m + 1$ and
$p^2 - 1 = 12(2m + 1)(6m + 4) = 24(2m + 1)(3m + 2)$and is divisible by 24.

$$p = 6n - 1 \Rightarrow p^2 - 1 = (6n - 1) - 1 = 36n^2 - 12n + 1 - 1$$
$$= 36n^2 - 12n = 12n(3n - 1)$$

If n is even then $n = 2m$ and $p^2 - 1 = 24m(6m - 1)$ and is divisible by 24.

If n is odd then $n = 2m + 1$ and
$p^2 - 1 = 24(2m + 1)(3m + 1)$ and is divisible by 24.

In each $p^2 - 1$ case is divisible by 24.

Q4 Suppose that n is odd
$\Rightarrow n = 2m + 1$
$\Rightarrow n^3 = (2m + 1)^3 = 8m^3 + 12m^2 + 6m + 1$
$= 2m(4m^2 + 6m + 3) + 1$

So n^3 is odd.

The cube of an odd number is always odd i.e. it cannot be even. So if n^3 is even n cannot be odd, it must be even.

You are given that n^3 is even and you must prove that n is even. Start by supposing that n is odd and use the method of proof by contradiction.

Q5 Suppose that the cube root of 2 is a rational number. $\sqrt[3]{2} = \frac{p}{q}$ where p and q are integers with no common factors. Cube both sides.

$2 = \frac{p^3}{q^3}$

Assume that the statement you wish to prove is false.

$p^3 = 2q^3 \Rightarrow p^3$ is even $\Rightarrow p$ is even $\Rightarrow p = 2m$

$\Rightarrow 2q^3 = (2m)^3 = 8m^3 \Rightarrow q^3 = 4m^3 \Rightarrow q$ is even

At this stage you need to find another result. If p^3 is even what can you say about p? You need to prove that p is even (you did this in Q4). This is an example of the proof of one result requiring the proof of another.

$\Rightarrow p$ and q have a common factor of 2. This is a contradiction so $\sqrt[3]{2}$ is not a rational number.

At the start of the proof you said that p and q have no common factor. This is the contradiction.

Q6

(a) Every odd number can be written as $2m + 1$ for some integer m.

m is even $\Rightarrow m = 2n \Rightarrow 2m + 1 = 4n + 1$

m is odd $\Rightarrow m = 2n + 1 \Rightarrow 2m + 1 = 4n + 3$

Every odd number can be written as $4n + 1$ or $4n + 3$.

(b) Consider two integers $4p + 1$ and $4q + 1$.

$(4p + 1)(4q + 1) = 16pq + 4p + 4q + 1 = 4(4pq + p + q) + 1$
$= 4k + 1$
where $k = 4pq + p + q$ is an integer.
The product is of the form $4n + 1$.

Start with two integers of the form $4n + 1$.

(c) The converse statement is:
'Any integer of the form $4n + 1$ can be written as a product of two integers of the form $4n + 1$ which are both greater than 1.'

State the converse statement.

A counter example is the integer 21, which can be written as $4 \times 5 + 1$, but neither of the factors 7 and 3 can be written in the form $4n + 1$.

There are many counter-examples.

7 Coordinate geometry

Q1

$x^2 + y^2 = 6x$
$x^2 + y^2 - 6x = 0$
$(x-3)^2 + y^2 - 9 = 0$

The coordinates of the centre of the circle are (3, 0) and the radius is 3.

Rewrite the equation in standard form to find the centre and radius.

$y = k - x$
$(x-3)^2 + (k-x)^2 - 9 = 0$
$2x^2 - x(6 + 2k) + k^2 = 0$
For a tangent: $(6 + 2k)^2 - 8k^2 = 0$
$6 + 2k = \pm\sqrt{8}k = \pm 2\sqrt{2}k$

$k = \frac{3}{\sqrt{2}-1}$ or $k = -\frac{3}{\sqrt{2}-1}$

Substitute $y = k - x$ in the equation and remember that if the line is a tangent it touches the circle once. This means that the quadratic in x has only one solution so '$b^2 - 4ac$' = 0, using the quadratic formula for solving the equation. Leave the answers in surd form.

Q2

$x^2 + y^2 + 4x - 10y + 13 = 0$

$(x+2)^2 + (y-5)^2 - 16 = 0$

The centre of the circle is (–2, 5) and its radius is 4.

$CP^2 = (2 - (-2))^2 + (3-5)^2 = 20$

$CP = \sqrt{20}$

Draw a sketch diagram showing the centre C, point P and tangents PT_1 and PT_2. Then you can see that you can use Pythagoras' theorem to give $PT^2 + r^2 = PC^2$.

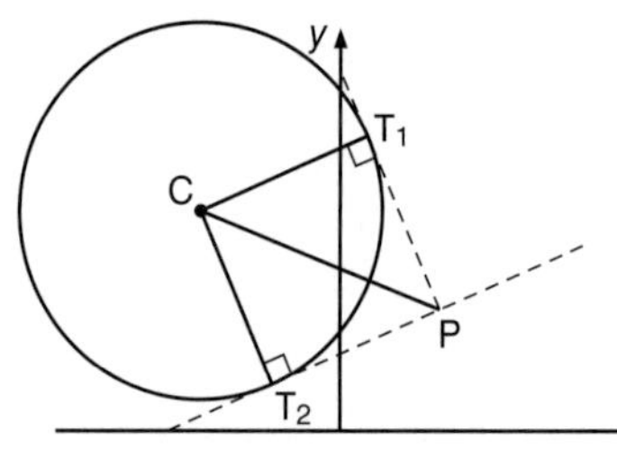

$PT^2 = PC^2 - r^2 = 20 - 16 = 4$

$PT = 2$

Q3

$x^2 + y^2 - 6x - 8y + 24 = 0$
$x^2 + m^2x^2 - 6x - 8mx + 24 = 0$
$(1 + m^2)x^2 - 2x(3 + 4m) + 24 = 0$

Substitute $y = mx$ and simplify the expression.

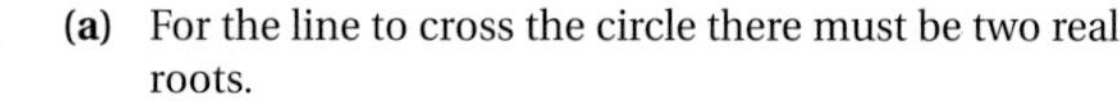

(a) For the line to cross the circle there must be two real roots.

Remember the discriminant of the quadratic formula. The number of solutions for x depends on whether the roots are real ($b^2 - 4ac > 0$), equal ($b^2 - 4ac = 0$) or complex ($b^2 - 4ac < 0$).

Use your graphic calculator to draw the quadratic in m and interpret the graph for $f(m) > 0, = 0, < 0$.

$4(3 + 4m)^2 - 4(1 + m^2) \times 24 > 0$

$4(9 + 24m + 16m^2) - 4 \times 24(1 + m^2) > 0$

$9 + 24m + 16m^2 - 24(1 + m^2) > 0$

$-15 + 24m - 8m^2 > 0$

$$m = \frac{-24 \pm \sqrt{24^2 - 4 \times -15 \times -8}}{-16} = \frac{-6 \pm \sqrt{6}}{-4}$$

For two real roots:
$\frac{6-\sqrt{6}}{4} < m < \frac{6+\sqrt{6}}{4}$

(b) If the line is a tangent it touches the circle just once. The equation will have one root.
For one root $m = \frac{6-\sqrt{6}}{4}$ or $m = \frac{6+\sqrt{6}}{4}$.

(c) If the line misses the circle there are no real roots.
For no roots $m < \frac{6-\sqrt{6}}{4}$ or $m > \frac{6+\sqrt{6}}{4}$.

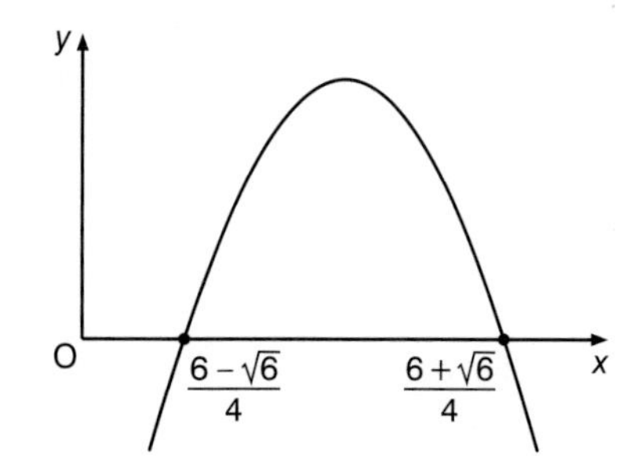

Answers	How to solve these questions

The centre is the midpoint of the line, with coordinates $(\frac{1}{2}(5 + (-3)), \frac{1}{2}(5 + (-1))) = (1, 2)$.

The coordinates of the midpoint of the line joining the two points $\left(\frac{x_1 + x_2}{2}, \frac{y_1 + y_2}{2}\right)$ give the centre.

The radius is $\sqrt{((5 - 1)^2 + (5 - 2)^2} = \sqrt{25} = 5$

Substitute in the formula $(x - a)^2 + (y - b)^2 = r^2$ to find the equation of the circle.

The equation of the circle is $(x - 1)^2 + (y - 2)^2 = 25$

$x^2 + y^2 - 2x - 4y - 20 = 0$

When the line $y = 3x - 16$ meets the circle:

$x^2 + (3x - 16)^2 - 2x - 4(3x - 16) - 20 = 0$

Substitute $y = 3x - 16$ to find where the line cuts the circle.

$10x^2 - 110x + 300 = 0 \Rightarrow x^2 - 11x + 30 = 0$

$\Rightarrow (x - 6)(x - 5) = 0$

The coordinates of P and Q are $(5, -1)$ and $(6, 2)$.

$x^2 + y^2 + 2x - 8y + 12 = 0$

Express the equation in the standard form to obtain the radius and centre.

$(x + 1)^2 + (y - 4)^2 - 17 + 12 = 0$

$(x + 1)^2 + (y - 4)^2 = 5$

So the circle has radius $\sqrt{5}$ and centre $(-1, 4)$.

Where the line $y = 2x + 1$ intersects the circle:

$x^2 + (2x + 1)^2 + 2x - 8(2x + 1) + 12 = 0$

$x^2 + 4x^2 + 4x + 1 + 2x - 16x - 8 + 12 = 0$

$5x^2 - 10x + 5 = 0 \Rightarrow x^2 - 2x + 1 = 0 \Rightarrow (x - 1)^2 = 0$

$x = 1, y = 3$

The line touches the circle as a tangent at the point $(1, 3)$.

For the rear wheel: centre $(0, 3)$ radius 3

Start by finding the centre and radius of each circle.

For the front wheel: centre $(14, 10)$ radius 10

(a) **(i)** The distance between the centres is $\sqrt{14^2 + (10 - 3)^2} = 7\sqrt{5}$ units

The distance in centimetres is 78.3 cm.

Remember to multiply by 5 to change from units on the graph to cm.

(ii) The smallest gap $= 35\sqrt{5} - 5 \times (3 + 10) = 35\sqrt{5} - 65 = 13.3$ cm

The smallest gap is the distance between the centres minus the sum of the radii.

(b) **(i)** The gradient of PB $= \frac{10 - 3}{14 - 7} = 1$

(ii) The gradient of AC $= -1$

The equation of AC is $\frac{y - 3}{x - 7} = -1$

$y = -x + 10$

$x^2 + (-x + 10)^2 - 28x - 20(-x + 10) + 196 = 0$

$2x^2 - 28x + 96 = 0$

$x^2 - 14x + 48 = 0$

$x = \frac{14 \pm \sqrt{14^2 - 4 \times 48}}{2} = \frac{14 \pm 2}{2}$

Find the gradient, then substitute known values in the equation for the circle and solve for x. You could factorise to give $(x - 6)(x - 8)$ instead of using the quadratic formula.

$x = 6$ or $x = 8$

$y = 4$ $y = 2$

A is $(8, 2)$ and C is $(6, 4)$.

$d = \sqrt{(2t - 1 - 8)^2 + (5t + 1 - 9)^2} = \sqrt{(2t - 9)^2 + (5t - 8)^2}$

$= \sqrt{29t^2 - 116t + 145}$

$= \sqrt{29(t - 2)^2 + 29}$

The distance will be given by $d = \sqrt{(x - 8)^2 + (y - 9)^2}$.

The minimum value of d is $\sqrt{29}$.

Q1

(a) $\frac{\mathrm{d}x}{\mathrm{d}t} = -kx,\ k > 0$

The question tells you that the rate of change of x $(\frac{\mathrm{d}x}{\mathrm{d}t})$ is decreasing and proportional to x. So rate of change is proportional to $-x$.

(b) $\int \frac{1}{x}\mathrm{d}x = -\int k\mathrm{d}t$

Separate the variables by dividing both sides by x.

$\ln x = -kt + c$

When $t = 0$, $x = x_0 \Rightarrow \ln x_0 = 0 + c \Rightarrow \ln x - \ln x_0 = -kt$

$\ln\frac{x}{x_0} = -kt$

Remember that '$x = x_0$ initially' means $x = x_0$ when $t = 0$. Substitute for $x = x_0$ when $t = 0$ to find c.

(c) When $t = 3$, $x = \frac{1}{2}x_0 \Rightarrow \ln\frac{1}{2} = -3k \Rightarrow k = \frac{1}{3}\ln 2$

Substitute $x = \frac{1}{2}x_0$, i.e. half the initial value when $t = 3$.

When $t = T$, $x = \frac{1}{10}x_0 \Rightarrow \ln\frac{1}{10} = -kT = -\frac{1}{3}\ln 2 \times T$

$T = 3\frac{\ln 10}{\ln 2} = 9.97$ minutes

$T = 10$ minutes to the nearest minute

(d) (i) $\frac{1}{3}\ln 2 = -0.231$ so $x = x_0 e^{-0.231t}$

Remember that if $\ln x = a$ then $x = e^a$.

(ii)

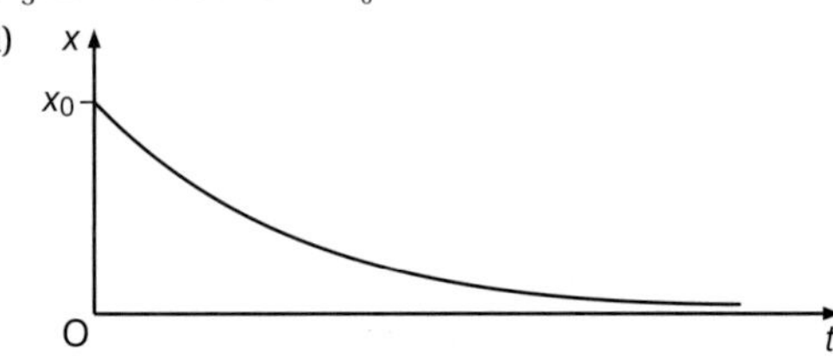

It is important to show that $x \to 0$ as t increases and $x = x_0$ when $t = 0$.

Q2

(a) Alan's model:

$\frac{\mathrm{d}h}{\mathrm{d}t} = c$

$h = ct + b$

When $t = 0$, $h = 10$, and when $t = 40$, $h = 5$.

$10 = 0 + b \Rightarrow b = 10$

Use the given conditions to find the values of the constants b and c.

$5 = 40c + 10 \Rightarrow c = -\frac{1}{8}$

$h = 10 - \frac{1}{8}t$

(b) Bhavana's model:

$\frac{\mathrm{d}h}{\mathrm{d}t} = kh$

Don't forget the constant of proportionality k.

$\int \frac{1}{h}\mathrm{d}h = \int k\mathrm{d}t$

$\ln h = kt + a$

At this stage you could write $h = Ae^{kt}$.

$\ln 10 = 0 + a \rightarrow a = \ln 10$

Use the given conditions to find the values of the constants a and k.

$\ln 5 = 40k + \ln 10 \Rightarrow k = -\frac{1}{40}\ln 2$

$\ln h = (-\frac{1}{40}\ln 2)t + \ln 10$

Remember that $\ln 5 - \ln 10 = \ln\frac{5}{10} = \ln\frac{1}{2} = -\ln 2$.

$h = 10e^{(-\frac{1}{40}\ln 2)t} = 10e^{-0.017t}$

(c) When $t = 60$:

Substitute $t = 60$ to find h for each model.

Alan's model gives $h = 10 - \frac{60}{8} = 2.5$

Bhavana's model gives $h = 10e^{-\frac{60}{40}\ln 2} = 3.5$

(d) Bhavana's model is better because the tank is likely to leak more slowly as the depth decreases.

Answers / How to solve these questions

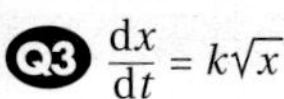

Q3 $\frac{dx}{dt} = k\sqrt{x}$

You are told that the rate is proportional to $\sqrt{x}$. You need to find the time for which the tank has been leaking so assume that initially $x = 200$. Call this time T.

When $t = 0$, $x = 200$

When $t = T$, $x = 100$ and $\frac{dx}{dt} = -1$

$-1 = k\sqrt{100} = 10k$

Substitute $x = 100$, $\frac{dx}{dt} = -1$ to find k.

$k = -\frac{1}{10}$

$\frac{dx}{dt} = -\frac{1}{10}\sqrt{x}$

Separate the variables by dividing both sides by $\sqrt{x}$.

$\int \frac{1}{\sqrt{x}} dx = -\frac{1}{10}\int dt$

$2\sqrt{x} = -\frac{t}{10} + c$

$2\sqrt{100} = 0 + c$

$c = 20\sqrt{2}$

Use the initial condition to find the value of c.

$\sqrt{x} = -\frac{t}{20} + 10\sqrt{2}$

Use $x = 100$ when $t = T$ to give an equation for T.

$\sqrt{100} = -\frac{T}{20} + 10\sqrt{2}$

$T = 20 \times 10(\sqrt{2} - 1) = 82.8$

The tank has been leaking for 82.8 minutes (which is about 80 minutes to the nearest 10 minutes).

Give your final answer in words.

Q4 $x^2\frac{dy}{dx} = y + 1$

First separate the variables by collecting all the ys on the left-hand side and all the xs on the right-hand side. Remember to divide both sides by $y + 1$ and both sides by x^2.

$\int \frac{1}{y+1} dy = \int \frac{1}{x^2} dx$

$\ln|y + 1| = -\frac{1}{x} + c$

When $x = 1$, $y = 0$

Substitute for the given values to find the value of c.

$\ln|1| = -1 + c$

$c = 1$

$\ln|y + 1| = 1 - \frac{1}{x}$

Rearrange the equation to write y as a function of x.

$y + 1 = e^{1 - \frac{1}{x}}$

$y = e^{1 - \frac{1}{x}} - 1$

Q5 $\frac{dv}{dt} = 10 - 3v$

$\int \frac{1}{10 - 3v} dv = \int dt$

Remember to divide both sides by $(10 - 3v)$. This separates the variables. (Note that if you add $3v$ to both sides you cannot proceed.)

$-\frac{1}{3}\ln|10 - 3v| = t + c$

Integrate both sides.

When $t = 0$, $v = 2$

Substitute the given conditions.

$-\frac{1}{3}\ln 4 = 0 + c \Rightarrow c = -\frac{1}{3}\ln 4$

$-\frac{1}{3}\ln|10 - 3v| = t - \frac{1}{3}\ln 4$

$\ln\left|\frac{10 - 3v}{4}\right| = -3t$

$\frac{10 - 3v}{4} = e^{-3t}$

$v = \frac{10}{3} - \frac{4}{3}e^{-3t}$

Rearrange the solution to find v in terms of t.

As $t \to \infty$ $v \to \frac{10}{3}$ since $e^{-3t} \to 0$

The limiting value of v occurs as $t \to \infty$.

Q6

(a)

n	t	$\frac{dn}{dt}$	dt	dn
10	0	9.98	5	49.9
59.9	5	59.2	5	296
356	10			

To calculate dn you multiply by 5 (dt) and to calculate the new value of n you add the current value of n to dn.

(b) The estimate can be improved by choosing a smaller step size for dt.

9 Partial fractions and integration

$$\frac{x}{(x+2)(x+3)} = \frac{A}{x+2} + \frac{B}{x+3}$$

First identify the form of partial fractions required.

$x = A(x+3) + B(x+2)$

$x = (A+B)x + (3A+2B)$

If the left- and right-hand sides have the same denominators, the two numerators must be equal.

$A + B = 1$ and $3A + 2B = 0$

Use this to form and then solve a pair of simultaneous equations.

$A = -2$ and $B = 3$

Alternatively, substitute $x = -2$ and $x = -3$ to find A and B.

Once you have found the values of A and B, you can tackle the integration.

$$\int_0^1 \frac{x}{(x+2)(x+3)}\,dx = \int_0^1 \frac{-2}{x+2}\,dx + \int_0^1 \frac{3}{x+3}\,dx$$

$$= \Big[-2\ln|x+2| + 3\ln|x+3|\Big]_0^1$$

You need to use the rules of logarithms to simplify the answer.

$= -2\ln3 + 3\ln4 - (-2\ln2 + 3\ln3)$
$= 8\ln2 - 5\ln3$

Note that $3\ln4 = 6\ln2$

Q2

$$\frac{15-13x+4x^2}{(1-x)^2(4-x)} = \frac{A}{1-x} + \frac{B}{(1-x)^2} + \frac{C}{4-x}$$

$$= \frac{A(1-x)(4-x) + B(4-x) + C(1-x)^2}{(1-x)^2(4-x)}$$

First decide which form of partial fraction is required, noting the squared term in the denominator.

$15 - 13x + 4x^2 = A(1-x)(4-x) + B(4-x) + C(1-x)^2$

If $x = 1$: $15 - 13 + 4 = 3B \Rightarrow B = 2$

If $x = 4$: $15 - 13 \times 4 + 4 \times 16 = 9C \Rightarrow C = 3$

Use the substitution method, with $x = 1$ and $x = 4$, to find B and C.

Compare coefficients of x^2: $4 = A + 3 \Rightarrow A = 1$

Compare coefficients to find A.

$$\int_2^3 \frac{15-13x+4x^2}{(1-x)^2(4-x)} = \int_2^3 \left(\frac{1}{1-x} + \frac{2}{(1-x)^2} + \frac{3}{4-x}\right)dx$$

$$= \left[-\ln|1-x| + \frac{2}{1-x} - 3\ln|4-x|\right]_2^3$$

Now tackle the integration.

$= (-\ln2 - 1 - 3\ln1) - (-\ln1 - 2 - 3\ln2)$

$= 1 + 2\ln2 = 1 + \ln4$

Use the laws of logarithms to simplify the answer and reach the required result.

(a) $\dfrac{A}{(x+1)^2} + \dfrac{B}{x+1} + \dfrac{C}{x+2}$

$$= \frac{A(x+2) + B(x+1)(x+2) + C(x+1)^2}{(x+1)^2(x+2)}$$

$$= \frac{(B+C)x^2 + (A+3B+2C)x + (2A+2B+C)}{(x+1)^2(x+2)}$$

First use a common denominator to bring the three fractions into a single fraction.

$x + 4 = (B+C)x^2 + (A+3B+2C)x + (2A+2B+C)$

Comparing coefficients of x^2:
$B + C = 0$ (1)
Comparing coefficients of x:
$A + 3B + 2C = 1$ (2)
Comparing constants:
$2A + 2B + C = 4$ (3)

Then compare coefficients to form three equations.

Solve these equations to find A, B and C.

$A + B = 1$ from (2) using $C = -B$
$2A + B = 4$ from (3) using $C = -B$
$A = 3$, $B = -2$, $C = 2$

Alternatively, use the substitution method to find A and C first.

(b) $f'(x) = \dfrac{-2\times3}{(x+1)^3} - \dfrac{-1\times2}{(x+1)^2} + \dfrac{-1\times2}{(x+2)^2}$

$f'(1) = -\frac{6}{8} + \frac{2}{4} - \frac{2}{9} = -\frac{17}{36}$

Differentiate each term and then substitute $x = 1$. Do not use decimal approximations when simplifying.

(c) $$\int_0^3 f(x)\,dx = \int_0^3 \left(\frac{3}{(x+1)^2} - \frac{2}{x+1} + \frac{2}{x+2}\right)dx$$

$$= \left[\frac{-3}{x+1} - 2\ln|x+1| + 2\ln|x+2|\right]_0^3$$

Integrate $f(x)$ using the partial fractions from part (a).

$= (-\frac{3}{4} - 2\ln4 + 2\ln5) - (-3 - 2\ln1 + 2\ln2)$

$= \frac{9}{4} + \ln\frac{25}{64}$

Simplify to the required form, noting that $\ln1 = 0$ and using the laws of logarithms to manipulate the answer as necessary.

Answers

How to solve these questions

Q4 $\dfrac{2-x-2x^2}{(1-x)(1-2x)^2}$

This question gives no clues how to form the partial fractions. Start by checking that the largest power on the top is less than the largest power on the bottom.

$$= \frac{A}{1-x} + \frac{B}{1-2x} + \frac{C}{(1-2x)^2}$$

The $(1-x)$ in the denominator requires the $\dfrac{A}{1-x}$ term and the $(1-2x)^2$ requires the $\dfrac{B}{1-2x} + \dfrac{C}{(1-2x)^2}$.

$$= \frac{A(1-2x)^2 + B(1-x)(1-2x) + C(1-x)}{(1-x)(1-2x)^2}$$

$$= \frac{(4A+2B)x^2 + (-4A-3B-C)x + (A+B+C)}{(1-x)(1-2x)^2}$$

$2 - x - 2x^2 = (4A+2B)x^2 + (-4A-3B-C)x + (A+B+C)$

Comparing coefficients of x^2: $4A + 2B = -2$ (1)
Comparing coefficients of x: $-4A - 3B - C = -1$ (2)
Comparing constant terms: $A + B + C = 2$ (3)

Find the values of A, B and C by combining the fractions and comparing coefficients. Alternatively, use the substitution method to find A and C.

$B = 1 - 2A$
$-2A + C = 4$
$-A + C = 3$
$A = -1,\ B = 1,\ C = 2$

$$\frac{2-x-2x^2}{(1-x)(1-2x)^2} = \frac{-1}{1-x} + \frac{1}{1-2x} + \frac{2}{(1-2x)^2}$$

Q5

(a) $\dfrac{18}{x^2(x+3)} = \dfrac{A}{x} + \dfrac{B}{x^2} + \dfrac{C}{x+3}$

The form of the partial fraction is given in the question. Form a single fraction and compare coefficients with the original fraction to find the values of A, B and C. Alternatively, use the substitution method.

$$= \frac{Ax(x+3) + B(x+3) + Cx^2}{x^2(x+3)}$$

$$= \frac{(A+C)x^2 + (3A+B)x + 3B}{x^2(x+3)}$$

$18 = (A+C)x^2 + (3A+B)x + 3B$

Comparing coefficients of x^2: $A + C = 0$ (1)
Comparing coefficients of x: $3A + B = 0$ (2)
Comparing constant terms: $3B = 18$ (3)
$B = 6,\ A = -2,\ C = 2$

$$\frac{18}{x^2(x+3)} = \frac{-2}{x} + \frac{6}{x^2} + \frac{2}{x+3}$$

(b) $\displaystyle\int_1^3 \frac{18}{x^2(x+3)}\,dx$

Now tackle the integration.

$$= \int_1^3 \left(\frac{-2}{x} + \frac{6}{x^2} + \frac{2}{x+3}\right)dx$$

$$= \left[-2\ln x - \frac{6}{x} + 2\ln(x+3)\right]_1^3$$

$= (-2\ln 3 - 2 + 2\ln 6) - (-2\ln 1 - 6 + 2\ln 4)$

$= 4 + \ln\left(\dfrac{36}{9 \times 16}\right) = 4 - 2\ln 2$

You need to know the laws of logarithms to bring all the terms into one expression. Remember that $-2\ln 3 = \ln\frac{1}{9}$, $2\ln 6 = \ln 36$, $2\ln 1 = 0$ and $-2\ln 4 = \ln\frac{1}{16}$.

Q6 $\dfrac{4x+3}{x(x+1)^2} = \dfrac{A}{x} + \dfrac{B}{x+1} + \dfrac{C}{(x+1)^2}$

First, decide on the form of the partial fractions. You need both $\dfrac{B}{x+1}$ and $\dfrac{C}{(x+1)^2}$.

$$= \frac{A(x+1)^2 + Bx(x+1) + Cx}{x(x+1)^2}$$

$$= \frac{(A+B)x^2 + (2A+B+C)x + A}{x(x+1)^2}$$

$4x + 3 = (A+B)x^2 + (2A+B+C)x + A$

Comparing coefficients of x^2: $A + B = 0$ (1)
Comparing coefficients of x: $2A + B + C = 4$ (2)
Comparing constant terms: $A = 3$ (3)

Compare the coefficients to form three equations, then solve them to find the values of A, B and C. Alternatively, use the substitution method.

$A = 3,\ B = -3,\ C = 1$

$$\frac{4x+3}{x(x+1)^2} = \frac{3}{x} - \frac{3}{x+1} + \frac{1}{(x+1)^2}$$

$$\int_2^4 \frac{4x+3}{x(x+1)^2}\,dx = \int_2^4 \left(\frac{3}{x} - \frac{3}{x+1} + \frac{1}{(x+1)^2}\right)dx$$

Now tackle the integration.

$$= \left[3\ln x - 3\ln(x+1) - \frac{1}{(x+1)}\right]_2^4$$

$= (3\ln 4 - 3\ln 5 - \frac{1}{5}) - (3\ln 2 - 3\ln 3 - \frac{1}{3})$

$= \frac{2}{15} + \ln\frac{216}{125}$

Use the laws of logarithms to manipulate the result to achieve the final answer in its simplest form.

10 MOMENTS AND CENTRES OF MASS

 Taking moments about A:

$(20 + 10)\bar{x} = 0.7 \times 20 + 1.6 \times 10$

Take moments to find the distance of the centre of mass from A.

$\bar{x} = \frac{30}{30} = 1$

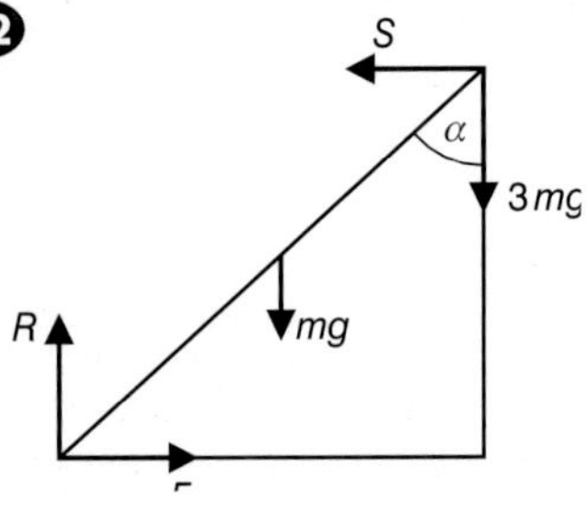

A diagram to show the forces acting is an important first step in a problem like this.

Resolving forces vertically:

$R = mg + 3mg = 4mg$

Find the reaction force acting at the base of the ladder.

Taking moments about the bottom of the ladder:

$2a\cos\alpha \times S = 2a\sin\alpha \times 3mg + a\sin\alpha \times mg$

You can take moments about either the top or the bottom of the ladder. The bottom of the ladder has been used in this case.

$S = \dfrac{7mg\sin\alpha}{2\cos\alpha} = \dfrac{7mg\tan\alpha}{2}$

$F = S = \dfrac{7mg\tan\alpha}{2}$

Note that as the ladder is in equilibrium, $F = S$.

$F \leqslant \mu R$

$\dfrac{7mg\tan\alpha}{2} \leqslant \frac{1}{4} \times 4mg$

$\tan\alpha \leqslant \frac{2}{7}$

Finally use the inequality $F \leqslant \mu R$ and solve for $\tan\alpha$.

Q3

(a) Area of square ABDE $= 18^2 = 324\,\text{cm}^2$

Area of triangle BCD $= \frac{1}{2} \times 18 \times 12 = 108\,\text{cm}^2$

Use Pythagoras' theorem to calculate the height of the triangle as 12 cm.

$(324 + 108)\bar{x} = 9 \times 324 + 22 \times 108$

$\bar{x} = \frac{5292}{432} = 12.25\,\text{cm}$

Calculate the position of the centre of mass using $\bar{x} = \dfrac{\Sigma m_i x_i}{\Sigma m_i}$ and assuming that the areas are proportional to the mass.

The centre of mass of the triangle is at a distance of $\frac{1}{3}$ of its height from the base. This result is in your formula book.

(b)

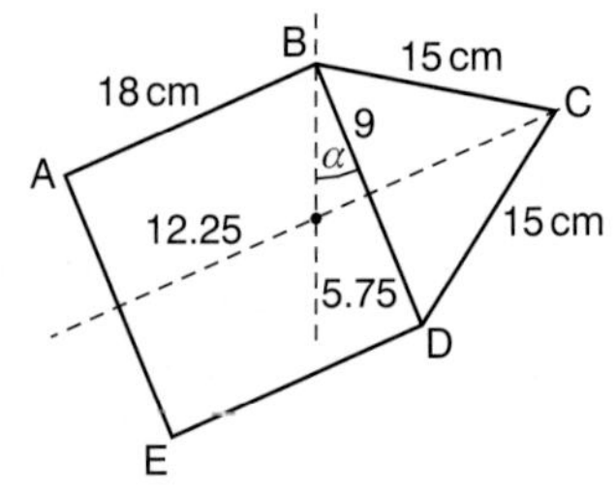

When the lamina is in equilibrium the centre of mass will be directly below the point of suspension. A clear diagram will help you to find the angle.

$\tan\alpha = \frac{5.75}{9}$

$\alpha = 32.6°$

Q4 $(40 + 48)\bar{x} = 40 \times 2 + 48 \times 7$

$\bar{x} = \frac{416}{88} = \frac{52}{11} = 4.73\,\text{cm}$

$(40 + 48)\bar{y} = 40 \times 5 + 48 \times 4$

$\bar{y} = \frac{392}{88} = \frac{49}{11} = 4.45\,\text{cm}$

For this problem you need to find the distance of the centre of mass from AF ($\bar{x}$) and the distance of the centre of mass from AB ($\bar{y}$).

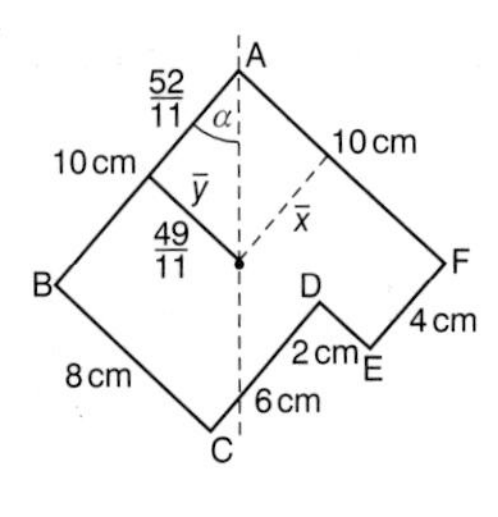

Once you know $\bar{x}$ and $\bar{y}$ you can draw a diagram, with the centre of mass below A, and calculate the required angle. A diagram will help you to make sure that you calculate the correct angle.

$\tan\alpha = \frac{49}{52}$

$\alpha = 43.3°$

Answers / How to solve these questions

Q5

(a) Weight of triangle = $\frac{W}{3}$

Weight of square = $\frac{2W}{3}$

$W\bar{x} = \frac{W}{3} \times \frac{2a}{3} + \frac{2W}{3} \times \frac{3a}{2}$

$\bar{x} = \frac{2a}{9} + a = \frac{11a}{9}$

The area of the triangle is half the area of the square. Use this fact to find the weight of each part.

The distance of the centre of mass of the triangle will be $\frac{2}{3}$ of the base of the triangle from A.

You can use these two parts of the lamina to find the position of the centre of mass.

(b) Taking moments about A:

$2a \times R_D = \frac{11a}{9} \times W$

$R_D = \frac{11W}{18}$

$R_D + R_A = W$

$R_A = \frac{7W}{18}$

You can take moments about either A or D. In this solution A has been used.

Use the fact that the two reactions are equal to the weight to find the other reaction force. Alternatively take moments about D and then check that the two reactions add to give W.

Q6

(a)

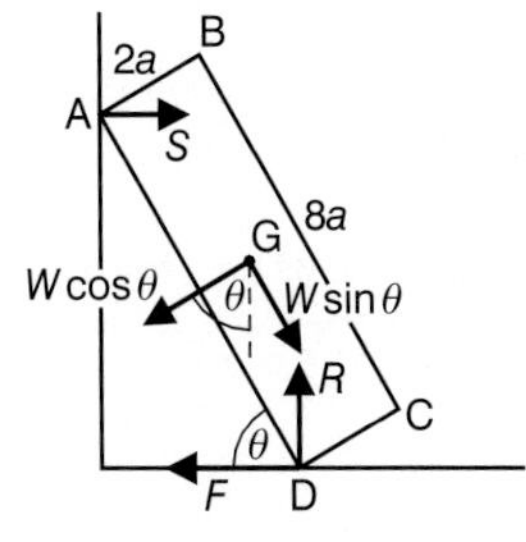

Anticlockwise moment about D = $W\cos\theta \times 4a - W\sin\theta \times a$

$= Wa(4\cos\theta - \sin\theta)$

Use a diagram to show the two components. Be careful to make sure that you don't mix up sin and cos in problems like this. Then take moments about D.

(b) In equilibrium, the moments are equal:

$Wa(4\cos\theta - \sin\theta) = S \times 8a\sin\theta$

$S = \frac{W}{8}\left(\frac{4}{\tan\theta} - 1\right)$

When the system is in equilibrium the anticlockwise moment of the weight will be balanced by a clockwise moment due to the reaction force, S, at the wall. This gives an equation that you can solve for S.

(c) On the point of slipping, $F = \mu R$

$F = S$

$R = W$

$\frac{W}{8}\left(\frac{4}{\tan\theta} - 1\right) = \mu W$

$4 - \tan\theta = 8\mu\tan\theta$

$4 = (8\mu + 1)\tan\theta$

$\tan\theta = \frac{4}{8\mu + 1}$

On the point of slipping $F = \mu R$. Note that $R = W$ and that $F = S$. Use these to form an equation, then solve it to find $\tan\theta$ as required.

Q7 On the point of sliding:

$R = mg\cos\alpha$

$F = mg\sin\alpha$

$F = \mu R$

$mg\sin\alpha = \mu mg\cos\alpha = 0.7mg\cos\alpha$

$\tan\alpha = 0.7 \Rightarrow \alpha = 35.0°$

First find the angle at which the block is in the point of sliding down the slope.

In this case use $F = \mu R$.

On the point of toppling:

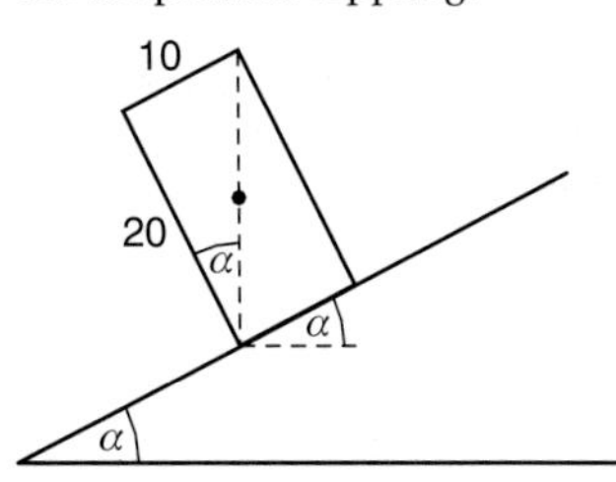

$\tan\alpha = \frac{1}{2} \Rightarrow \alpha = 26.6°$

The block topples before it slides.

Then find the angle at which the block would be on the point of toppling. In this case the centre of mass will be directly above the corner about which the block will topple. This is shown in the diagram.

State your conclusion to the problem clearly.

11 Energy

Q1 KE lost $= \frac{1}{2} \times 0.006 \times 400^2 - \frac{1}{2} \times 0.006 \times 250^2 = 292.5$ J

$0.02 \times F = 292.5$

$F = \frac{292.5}{0.02} = 14\,600$ N

Start by calculating the KE lost which will be equal to the work done against the resistance force. Then find the force, using work done = Fd.

Q2

(a) Loss of GPE $= 80 \times 9.8 \times 2$
$= 1568$ J

$1568 = \frac{1}{2} \times 80v^2 - \frac{1}{2} \times 80 \times 2^2$

$v = 6.57\ \text{m s}^{-1}$

First calculate the potential energy that is lost. This will be equal to the gain in kinetic energy and so you can calculate the value of v.

There is no air resistance.

Make sure that you include a valid assumption and state it clearly. One or two words are often enough.

(b) Maximum KE = 1728 J

$1728 = 80 \times 9.8h$

$h = \frac{1728}{80 \times 9.8} = 2.20$ m

Consider the kinetic energy of the soldier at the lowest point. This will be equal to the potential energy at C. Use this to calculate h.

(c) The tension acts perpendicular to the motion and so does not change the KE.

Q3

(a) KE lost $= \frac{1}{2} \times 1200 \times 30^2 - \frac{1}{2} \times 1200 \times 15^2$
$= 405\,000$ J

(b) $F = 0.8 \times 1200 \times 9.8$
$= 9408$ N

Work done against friction $= 9408 \times 25$
$= 235\,200$ J

Calculate the friction force using $F = \mu R$ as the car skids on a horizontal surface. Then multiply this by the distance to give the work done against friction.

(c) $25R = 405\,000 - 235\,200$
$R = 6792$ N

$(6792 + 9408)d = \frac{1}{2} \times 1200 \times 15^2$

$d = 8\frac{1}{3}$ m

Air resistance, R, must also account for the loss of some of the kinetic energy. Once you have found this, you can use the total of the resisting forces to find the further distance travelled.

Q4

(a) Work done by gravity $= 0.2 \times 9.8h = 1.96h$

Work done against resistance $= 0.8h$

$\frac{1}{2} \times 0.2 \times 3^2 = 1.96h - 0.8h$

$h = \frac{0.9}{1.16} = 0.776$ m

Consider the work done by gravity and the work done against the resistance force, in terms of h. The difference between these will give the final kinetic energy of the ball. Use this to find h.

(b) $0.2 \times 9.8 \times 0.3 + 0.8 \times 0.3 = \frac{1}{2} \times 0.2v^2$
$v = \sqrt{8.28} = 2.88\ \text{m s}^{-1}$

Work is done against both gravity and the resistance force. This must be equal to the kinetic energy of the ball when it leaves the ground.

Q5

(a) $F = 2000 + 4800 \times 9.8 \times \frac{1}{20}$
$= 4352$ N

$P = 4352 \times 12 = 52.2$ kW

A force must act to balance the component of the weight of the lorry parallel to the slope and the resistance force. Multiplying this by the speed gives the power.

(b) $4800a = 4352 - 2000$
$a = 0.49\ \text{m s}^{-2}$

A forward force of 4352 N will continue to act. Subtract the resistance force from this and apply Newton's second law to find the acceleration.

(c) $52\,224 = 2000v$
$v = 26.1\ \text{m s}^{-1}$

Use $P = Fv$ with $F = 2000$ to find the speed.

Answers	How to solve these questions

(a) Gain in KE $= \frac{1}{2} \times 650 \times 35^2 - \frac{1}{2} \times 650 \times 15^2$

$= 325\,000$ J

Work done by gravity $= 650 \times 9.8 \times 400\sin 5°$
$= 222\,073$ J

Work done by car $= 325\,000 - 222\,073$
$= 102\,927$ J

$102\,927 = 400F$

$F = 257$ N

The difference between the gain in kinetic energy and the work done by gravity will give the work done by the engine. Divided this by the distance travelled to find the magnitude of the force acting.

(b) $F = 900 - 650 \times 9.8\sin 5°$

$P = (900 - 650 \times 9.8\sin 5°) \times 35$
$= 12.1$ kW

First find the driving force, which will be the difference between the component of the weight parallel to the slope and the resistance force. Then multiply by the speed to get the power.

(c) $650a = F + 650 \times 9.8\sin 5° - 900$

$F = 650a - 650 \times 9.8\sin 5° + 900$

$12\,069 = 15(650a - 650 \times 9.8\sin 5° + 900)$

$a = 0.707 \text{ m s}^{-2}$

Express the driving force in terms of the acceleration, a. Then use $P = Fv$ to form an equation, and solve it for a.

Q7

(a) $R = kv$
$48\,000 = 40k \times 40$
$k = 30$
$R = 30v$

At its top speed the resistance force will be $40k$. Multiply this by the speed to give the maximum power and solve the equation for k.

(b) $F = 1200 \times 9.8\sin 4° + 30v$

$48\,000 = (1200 \times 9.8\sin 4° + 30v)v$

Find the force that is needed to keep the car moving up the slope at a constant speed.

$0 = 30v^2 + 1200 \times 9.8\sin 4° v - 48\,000$

Multiply this by v to find the power and form a quadratic equation. Solve it to find v.

$v = 28.6$ or -55.9

$v = 28.6 \text{ m s}^{-1}$

Note that the negative root must be rejected as this indicates that the car is travelling downhill.

12 Calculus in mechanics

$v = t^2(6 - t)$

Differentiate the velocity with respect to t to get the acceleration.

$a = 12t - 3t^2$

$a = 0$ when $12t - 3t^2 = 0$

$\Rightarrow t = 0$ or $t = 4$

Then solve the equation $a = 0$ to find two values for t, one of which is $t = 4$. Alternatively substitute $t = 4$ into the expression for the acceleration.

The particle is at rest when $v = 0$.

$t^2(6 - t) = 0 \Rightarrow t = 0, t = 6$

Solve the equation $t^2(6 - t) = 0$ to show that the velocity is zero when $t = 6$.

The particle comes to rest when $t = 6$ s.

$s = \int_0^6 (6t^2 - t^3)\,dt = \left[2t^3 - \frac{1}{4}t^4\right]_0^6 = 432 - 324$

$= 108$ m

Integrate the velocity between limits of integration of 0 and 6 to find the displacement.

(a) $\mathbf{r} = (t^3 - 3t)\mathbf{i} + 4t^2\mathbf{j}$

$\mathbf{v} = \frac{d\mathbf{r}}{dt} = (3t^2 - 3)\mathbf{i} + 8t\mathbf{j}$

Differentiate the position vector to get the velocity.

(b) When the particle is moving parallel to $\mathbf{i} + \mathbf{j}$:

$3t^2 - 3 = 8t$

$3t^2 - 8t - 3 = 0$

$(3t + 1)(t - 3) = 0$

$t = -\frac{1}{3}$ or $t = 3$

When the velocity is parallel to $\mathbf{i} + \mathbf{j}$, the two components of the velocity will be equal. Use this to form a quadratic equation, then solve it by factorising.

The particle is moving parallel to $\mathbf{i} + \mathbf{j}$ when $t = 3$.

As the question states that $t \geqslant 0$, only the positive solution applies.

Answers	How to solve these questions
Q3 $\mathbf{r} = \left(5t - \frac{t^2}{100}\right)\mathbf{i} + \left(3t + \frac{t^2}{20}\right)\mathbf{j}$	
$\mathbf{v} = \left(5 - \frac{t}{50}\right)\mathbf{i} + \left(3 + \frac{t}{10}\right)\mathbf{j}$	Differentiate the displacement to find the velocity.
When the particle is travelling due north:	
$5 - \frac{t}{50} = 0$ $t = 250$	The easterly component of the velocity will be zero when the particle is heading north. Use this to form an equation to find t.
$\mathbf{r} = \left(5 \times 250 - \frac{250^2}{100}\right)\mathbf{i} + \left(3 \times 25 + \frac{250^2}{100}\right)\mathbf{j}$ $= 625\mathbf{i} + 3875\mathbf{j}$	Substitute this value of t into the position vector.
Q4 $\mathbf{a} = t\mathbf{i} + \frac{2t}{5}\mathbf{j}$	
$\mathbf{v} = \int t\,dt\mathbf{i} + \int \frac{2t}{5}dt\mathbf{j} = \left(\frac{t^2}{2} + c_1\right)\mathbf{i} + \left(\frac{t^2}{5} + c_2\right)\mathbf{j}$	First integrate the acceleration to find the velocity.
$\mathbf{v} = 4\mathbf{i}$ when $t = 0 \Rightarrow c_1 = 4,\ c_2 = 0$	Use the initial velocity to find the constants of integration.
$\mathbf{r} = \int (\frac{t^2}{2} + 4)dt\mathbf{i} + \int \frac{t^2}{5}dt\mathbf{j}$ $= \left(\frac{t^3}{6} + 4t + c_3\right)\mathbf{i} + \left(\frac{t^3}{15} + c_4\right)\mathbf{j}$	Integrate the velocity to find the position vector.
$\mathbf{r} = 0\mathbf{i} + 0\mathbf{j}$ when $t = 0 \Rightarrow c_3 = 0,\ c_4 = 0$	Use the initial position to find the constants of integration.
$\frac{t^3}{15} = 1.8 \Rightarrow t = 3$	Use the northerly component of the position vector to find t.
If $t = 3$, $\frac{t^3}{6} + 4t = \frac{3^3}{6} + 4 \times 3$ $= 16.5$	Check that this value of t gives the correct **i** component of the position vector.
Q5	
(a) $\mathbf{a} = \frac{t}{2}\mathbf{i} + 2\mathbf{j}$	
$\mathbf{v} = \int \frac{t}{2}dt\mathbf{i} + \int 2dt\mathbf{j} = \left(\frac{t^2}{4} + c_1\right)\mathbf{i} + (2t + c_2)\mathbf{j}$	Integrate the acceleration to find the velocity.
From the initial velocity, $c_1 = 2$ and $c_2 = -6$. $\mathbf{v} = \left(\frac{t^2}{4} + 2\right)\mathbf{i} + (2t - 6)\mathbf{j}$	Use the initial velocity to find the constants of integration.
$\mathbf{r} = \int \left(\frac{t^2}{4} + 2\right)dt\mathbf{i} + \int (2t - 6)dt\mathbf{j}$ $= \left(\frac{t^3}{12} + 2t + c_3\right)\mathbf{i} + (t^2 - 6t + c_4)\mathbf{j}$	Integrate the velocity to find the position vector.
From the original position, $c_3 = 0$ and $c_4 = 4$. $\mathbf{r} = \left(\frac{t^3}{12} + 2t\right)\mathbf{i} + (t^2 - 6t + 4)\mathbf{j}$	Use the initial position to find the constants of integration.
(b) At $t = 2$, $\mathbf{r} = (\frac{8}{12} + 4)\mathbf{i} + (4 - 12 + 4)\mathbf{j} = \frac{14}{3}\mathbf{i} - 4\mathbf{j}$ Distance $= \sqrt{(\frac{14}{3})^2 + 4^2} = 6.14$ m	Substitute $t = 2$ to find the position and the distance from the starting point.
(c) At $t = 1$, $\mathbf{a} = \frac{1}{2}\mathbf{i} + 2\mathbf{j}$ $\mathbf{F} = 20(\frac{1}{2}\mathbf{i} + 2\mathbf{j}) = 10\mathbf{i} + 40\mathbf{j}$ $F = \sqrt{10^2 + 40^2} = 41.23$ N	First find the acceleration when $t = 1$. Then multiply the acceleration by the mass to obtain the force ($F = ma$).
Q6	
(a) Model A: $t = 0$ gives $\mathbf{r} = 0\mathbf{i} + 0\mathbf{j}$ $t = 2$ gives $\mathbf{r} = 10\mathbf{i} + 5\mathbf{j}$ Model B gives the same values.	Substitute $t = 0$ and $t = 2$ into each model.
(b) Model A: $\mathbf{v}_A = 5\mathbf{i} + \frac{5}{2}\mathbf{j}$	Differentiate to find the velocity of A.
The velocity is the same at all times $\Rightarrow$ zero acceleration.	Note that because this velocity is constant the acceleration must be zero.
Model B: $\mathbf{r}_B = \frac{5}{2}(3t^2 - t^3)\mathbf{i} + \frac{5}{4}(3t^2 - t^3)\mathbf{j}$ $\mathbf{v}_B = \frac{5}{2}(6t - 3t^2)\mathbf{i} + \frac{5}{4}(6t - 3t^2)\mathbf{j}$	Differentiate to find the velocity of B.
When $t = 0$, $\mathbf{v} = 0\mathbf{i} + 0\mathbf{j}$ When $t = 2$, $\mathbf{v} = 0\mathbf{i} + 0\mathbf{j}$	You need to substitute $t = 0$ and $t = 2$ as the velocity is not constant.

Answers	How to solve these questions

(c) Model B:
$\mathbf{a} = \frac{15}{2}(3 - 2t)\mathbf{i} + \frac{15}{4}(3 - 2t)\mathbf{j}$

Differentiate the velocity to get the acceleration.

There is zero acceleration when $t = 1.5$.

(d) Model **B** is better because with it the bird starts and ends at rest at each tree speeding up and slowing down as it flies between them.

Make sure that you state which you think is better and why.

Q7 $\mathbf{v} = (4 - 5\cos t)\mathbf{i} + (6 + 3\sin t)\mathbf{j}$

When the particle is travelling horizontally:

$$4 - 5\cos t = 0$$
$$\cos t = \frac{4}{5} = 0.8$$
$$t = 0.64\text{ s}$$

Differentiate the position vector to get the velocity. It will travel vertically when the horizontal component of the velocity is zero.

Don't forget to work in radians.

13 Projectiles

Q1

(a) At the highest point:
$26\sin\alpha - 9.8t = 0$

$$t = \frac{26\sin\alpha}{9.8} = \frac{26}{9.8} \times \frac{5}{13} = \frac{100}{98}$$

Begin by using the fact that the vertical component of the velocity will be zero at the highest point. Then use this to calculate the maximum height.

$$h = 26\sin\alpha \times \frac{100}{98} - \frac{9.6}{2} \times \frac{5}{13} \times \left(\frac{100}{98}\right)^2 + 0.8$$

Don't forget to include the initial height of 0.8 m. Note that $\sin\alpha = \frac{5}{13}$.

$$= 26 \times \frac{5}{13} \times \frac{100}{98} - 4.9\left(\frac{100}{98}\right)^2 + 0.8 = 5.9\text{ m}$$

(b) Horizontal distance = $V\cos\alpha \times t$

$$V\cos\alpha \times t = 36 \Rightarrow t = \frac{36}{26\cos\alpha} = \frac{36}{26} \times \frac{13}{12} = \frac{3}{2}$$

The horizontal distance travelled is given by $V\cos\alpha \times t$.
Use this to find t given that the distance is 36 m.

$$h = 26 \times \frac{5}{13} \times \frac{3}{2} - 4.9\left(\frac{3}{2}\right)^2 + 0.8 = 4.8\text{ m}$$

Use this value of t to find the height at which the ball hits the window.
Note that $\cos\alpha = \frac{12}{13}$.

(c) Air resistance

State one relevant factor.

Q2

(a) **(i)** $4.6 = 8\cos 40° \times t$

$$t = \frac{4.6}{8\cos 40°} = 0.75\text{ s}$$

Use the fact that the horizontal displacement is 4.6 m to find the time for the ball to be level with the centre of the net.

(ii) $h = 8\sin 40° \times 0.7506 - 4.9 \times 0.7506^2$
$= 1.1\text{ m}$

So the ball is 0.3 m below the basket.

Use the time you have already calculated to find the height of the ball relative to its starting point. As this is less than 1.4 m the ball does not reach the net.

(b) **(i)** $V\cos 40° \times t = 4.6$

$$t = \frac{4.6}{V\cos 40°}$$

First express the time to travel 4.6 m horizontally in terms of V.

$$1.4 = V\sin 40° \times t - 4.9t^2$$
$$1.4 = 4.6\tan 40° - 4.9 \times \frac{4.6^2}{V^2\cos^2 40°}$$

Then form an equation for the vertical motion, using the fact that the ball rises 1.4 m. Substitute for t using the expression above and solve the resulting equation for V.

$$V = \sqrt{\frac{4.9 \times 4.6^2}{(4.6\tan 40° - 1.4)\cos^2 40°}}$$
$$= 8.5\text{ m s}^{-1}\text{ (2 s.f.)}$$

Finding the vertical component of the velocity:

$$v = 8.5\sin 40° - 9.8t = 8.5\sin 40° - 9.8 \times \frac{4.6}{8.5\cos 40°}$$
$$= -1.46$$

Consider the vertical component of the velocity of the ball.

As this is negative the ball is moving downwards.

So the ball is descending.

Don't forget to make the statement at the end of your solution.

Answers | **How to solve these questions**

Q3

(a) $40 = 28\cos\theta t$

$t = \dfrac{10}{7\cos\theta}$

Use the horizontal distance travelled to express t in terms of θ.

$0.5 = 28t\sin\theta - 4.9t^2$

$0.5 = 40\tan\theta - 4.9 \times \left(\dfrac{10}{7\cos\theta}\right)^2$

Write down an equation for the vertical motion in terms of t and θ. Then substitute for t in this equation.

$0.5 = 40\tan\theta - 10\sec^2\theta$

$0.5 = 40\tan\theta - 10(1 + \tan^2\theta)$

$0 = 10\tan^2\theta - 40\tan\theta + 10.5$

Simplify this equation and use $\sec^2\theta = 1 + \tan^2\theta$ to form a quadratic equation in $\tan\theta$.

$\tan\theta = 3.718$ or 0.282

$\theta = 74.9°$ or $15.7°$

Solve the quadratic to find two values and hence the two possible angles of projection.

(b) Use the smaller angle as this will give the most direct path.

The lower angle is the one that is more likely to have been used as the large angle would give a very high trajectory.

Q4

(a) $100 = 300\sin 30° t - 4.9t^2$

$0 = 4.9t^2 - 150t + 100$

$t = 0.6819$ or 29.93

$t = 29.9$ s

$BC = 300\cos 30° \times 29.93 - 1000 = 6780$ m

Consider the vertical motion of the shell, and form an equation in t. Solve this quadratic equation to find two values for t. The smaller of these is when the shell is at a height of 100 m for the first time, that is, as it is rising. The second is when it hits the ground and it is this value that you should give as your final answer. The horizontal distance can then be calculated and 1000 m subtracted to give BC.

(b) $1000 = 300t\cos\theta$

$t = \dfrac{10}{3\cos\theta}$

$100 = 300t\sin\theta - 4.9t^2$

$100 = 1000\tan\theta - 4.9 \times \left(\dfrac{10}{3\cos\theta}\right)^2$

$900 = 9000\tan\theta - 490\sec^2\theta$

$90 = 900\tan\theta - 49(1 + \tan^2\theta)$

$0 = 49\tan^2\theta - 900\tan\theta + 139$

$\tan\theta = 18.21$ or 0.1558

This part of the problem requires you to form a quadratic equation for $\tan\theta$. Do this in the same way as in question 3, by using an equation for the horizontal displacement to get an expression for t and then substituting this into an equation for the vertical displacement.

$\theta = 86.9°$ or $8.9°$

$\theta = 8.9°$

You need to decide which value of θ to give as the answer. It must be the smaller value, because for the larger one the shell would be on its way down, rather than on its way up.

Q5

(a) There is no air resistance.

State one relevant reason.

(b) $20 = 28t\cos\theta \Rightarrow t = \dfrac{5}{7\cos\theta}$

Use the horizontal displacement to express t in terms of θ.

$27.5 < 28t\sin\theta - 4.9t^2$

$27.5 < 20\tan\theta - 2.5(1 + \tan^2\theta)$

Substitute this into the inequality for the vertical motion and use $\sec^2\theta = 1 + \tan^2\theta$ to form a quadratic in $\tan\theta$.

$\tan^2\theta - 8\tan\theta + 12 < 0$

$2 < \tan\theta < 6$

$63.4° < \theta < 80.5°$

Solve the quadratic to find the values of θ to give the range of values.

(c) When the ball lands:

$0 = 28t\sin\theta - 4.9t^2$

The ball will land when the vertical displacement is zero.

$t = 0$ or $t = \dfrac{28\sin\theta}{4.9}$

This will give the time of flight.

$R = 28\cos\theta \times \dfrac{28\sin\theta}{4.9}$

$= 80\sin 2\theta$

The time of flight can be used to find the horizontal displacement of the range. Use the trigonometric identity $\sin 2\theta = 2\sin\theta\cos\theta$.

(d) **(i)** Require smallest value of $\sin\theta$ so $\theta = 63.4°$.

(ii) Require largest value of $\sin 2\theta$ so $\theta = 63.4°$.

Make sure that you give reasons to support your answers.

14 Momentum and the coefficient of restitution

Q1

(a) $u = \sqrt{2g \times 1.2} = \sqrt{2.4g}$
$v = -\sqrt{2g \times 0.8} = -\sqrt{1.6g}$

First find the velocities before and after the bounce. This can be done with the constant acceleration equation $v^2 = u^2 + 2as$.

$-\sqrt{1.6g} = -e\sqrt{2.4g}$

Then use $v = -eu$ to find e.

$e = \dfrac{\sqrt{1.6}}{\sqrt{2.4}} = 0.816$

(b) $I = 0.25 \times (-\sqrt{1.6g}) - 0.25 \times (\sqrt{2.4g})$
$= -2.20$

Calculate the impulse using $I = mv - mu$, being careful with the signs of the velocities.

With air resistance, the speed before the bounce is less and the speed after the bounce is greater, so e would be larger.

Make sure that you give both a reason and a conclusion.

Q2

(a) $3mU = 3mv_A + mv_B$
$v_A - v_B = -U$
$v_A = v_B - U$
$3mU = 3mv_B - 3mU + mv_B$
$4mv_B = 6mU$

Use conservation of momentum to form one equation, and the coefficient of restitution to form a second equation. Note that in this case $e = 1$, as the collision is perfectly elastic.

$v_B = \dfrac{3U}{2}$

Solve the equations to give the required value for v_B.

(b) $v_A = \dfrac{3U}{2} - U = \dfrac{U}{2}$

Use the equations from part (a) to find v_A.

$-\dfrac{U}{2} = -e\dfrac{3U}{2}$

Then use the equation $v = -eu$ to find e.

$e = \frac{1}{3}$

Q3 $2m \times 3 + m \times (-3) = mv_P$

$v_P = 3 \text{ m s}^{-1}$

Define the masses of the particles as $2m$ and m. Then use conservation of momentum to find the velocity of P after the collision.

$3 - 0 = -e(-3 - 3)$

$3 = 6e$

$e = \frac{1}{2}$

The coefficient of restitution can be found now, as you know all the velocities before and after the collision.

Q4

(a) $m \times 3u + 2m \times (-u) = mv_A + 2mv_B$

$u = v_A + 2v_B$

Form equations using conservation of momentum and the coefficient of restitution, to find the velocity of B after the collision.

$v_A - v_B = -e(3u - (-u)) = -4eu$

$v_A = v_B - 4eu$

$u = v_B - 4eu + 2v_B$

$v_B = \frac{1}{3}(1 + 4e)u$

(b) $v_A = \frac{1}{3}(1 + 4e)u - 4eu$
$= \frac{1}{3}(1 - 8e)u$

Find the velocity of A after this collision. As A changes direction its velocity must be negative after the collision. This allows you to form and solve an inequality.

$v_A < 0$

$1 - 8e < 0$

$e > \frac{1}{8}$

(c) $v_B = -\frac{1}{2} \times \frac{1}{3}(1 + 4e)u$
$= -\frac{1}{6}(1 + 4e)u$

Speed of A < speed of B

Use $v = -eu$ to find the velocity of B after the collision with wall. Note that A and B are now both moving in the same direction. If they are to collide again the speed of B must be greater than the speed of A. Multiply both velocities by -1 to convert them into speeds before forming the inequality.

$\frac{1}{3}(8e - 1)u < \frac{1}{6}(1 + 4e)u$

$16e - 2 < 1 + 4e$

Alternatively work with the velocities, but solve the inequality $v_A > v_B$.

$12e < 3$

$e < \frac{1}{4}$

(a)

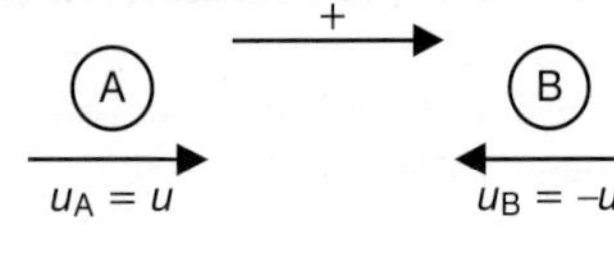

After (A) (B)

$v_A = -\frac{u}{2}$ $\quad$ v_B

Show the velocities clearly on the diagram. Be careful to give consideration to the velocities.

$$mu - Mu = -m\frac{u}{2} + Mv \Rightarrow Mv = \frac{3mu}{2} - Mu$$

To find the speed of B after the collision, use conservation of momentum.

$$v = u(\frac{3m}{2M} - 1)$$

As the direction of motion of B is reversed it will move in the positive direction after the collision. Hence you can use the inequality $v > 0$.

$$v > 0 \Rightarrow \frac{3m}{2M} - 1 > 0$$

$$\frac{m}{M} > \frac{2}{3}$$

(b) $-\frac{u}{2} - v = -e(u - (-u))$

Apply the equation involving the coefficient of restitution to form an expression. If you make x the subject of this expression, you can use it with $e \leqslant 1$ to find the first inequality.

$$e = \frac{1}{4} + \frac{v}{2u}$$

$$\frac{1}{4} + \frac{v}{2u} \leqslant 1 \Rightarrow v \leqslant \frac{3u}{2}$$

$$u(\frac{3m}{2M} - 1) \leqslant \frac{3u}{2} \Rightarrow \frac{3m}{2M} \leqslant \frac{5}{2}$$

To obtain the second inequality, substitute the expression for v obtained in part (a).

$$\frac{m}{M} \leqslant \frac{5}{3}$$

(c) $I = -(m + M)\frac{3u}{4} - (mu - Mu)$

$$= \frac{Mu}{4} - \frac{7mu}{4}$$

The impulse on B is the only external impulse on the whole system in the period under consideration. So the impulse on B will be the same as the impulse on the whole system. Hence the change in momentum of the system will be equal to the impulse on B.

15 The Poisson distribution

Q1 The Poisson can be used, with $\lambda = 3$:

$P(X < 5) = P(X \leqslant 4)$
$= 0.8153$ from distribution tables

However this is a 'rare event situation' since p is small and the mean, $np = 3$, is also small.

The situation described may be modelled using a binomial B(200, 0.015), which assumes that faults occur independently and that no computer is more or less likely to develop a fault than another.

Q2

(a) $P(X \geqslant 3) = 1 - P(X \leqslant 2)$
$= 1 - 0.0620 = 0.938$

This is a classic random events in a continuous medium context, hence the Poisson distribution, Poi(6), may be used.

(b) In the southern laboratory; P(0) is 0.606 53, hence the mean number of radioactive events in a second is $-\ln 0.606\,53 = 0.5000$.

Then the mean number in 6 seconds is 3. Hence:

$P(X \geqslant 3) = 1 - 0.4232 = 0.5768$

Use the uniform rate assumption required by Poisson distributions.

(c) The average number in a 15 second period in the individual laboratories is 1.5 + 7.5. Hence Poi(9) may be used.

$P(X = 6) = 0.2068 - 0.1157 = 0.091$

In this part, use the additive property of independent Poisson distributions.

Q3 $B(n, p)$ can be approximated by Poi(np). The probability, p, must be very small (e.g. usually sound for p no more than 0.02) and np must not be large e.g. 5 or less.

(a) Blood type occurs independently in the student body e.g. there are no identical twins. The likelihood of type AB– is constant across the student body e.g. there is no ethnic characteristic blood typing.

(b) B(300, 0.014) is the appropriate model. However, this may be approximated by Poi(4.2) since $p < 0.02$ and np is also small. Then the required probability is approximately:

$P(X \geqslant 5) = 1 - P(X < 5) = 1 - 0.5898 = 0.4102$

16 The chi-squared distribution

(a) The binomial must have $n = 5$ since there are five family members. Then the probabilities are: 0.55^5, $5 \times 0.45 \times 0.55^4, \ldots$

The value of 5.535 is 300 – the value of these frequencies.

Since there were 300 families, the expected frequencies are 300 times these probabilities: 15.099, 61.767, 101.073, 82.696, 33.830 and 5.535.

Give the E values to at least 1 d.p.

(b) H_0: binomial is appropriate, H_1: not appropriate

There are five degrees of freedom and the critical value is 11.070.

The test statistic is $((18 - 15.099)^2 \div 15.099) + \ldots)$
$= (0.5574 + 5.9888 + 2.556 + 1.9492 + 0.5140 + 1.0978)$
$= 12.6631$

This is greater than the critical value, hence the null hypothesis is rejected. The proposed binomial model is not appropriate.

(a) The mean number of errors is $132 \div 55 = 2.4$.

Here, the parameter is taken to be the mean.

(b) Expected frequencies = $55 \times P(0)$, $55 \times P(1), \ldots$

The final expected frequency is $55 - (5.0 + 12.0 + \ldots + 6.9) = 5.2$.

This gives $55 \times e^{-2.4}$, $55 \times e^{-2.4} \times 2.4, \ldots$ i.e. 5.0, 12.0, 14.4, 11.5, 6.9, then the next comes to 3.3 and the rest are smaller. These cells must be amalgamated so that the expected frequency is not less than 5.

No. errors	0	1	2	3	4	≥ 5
No. children	4	13	13	14	7	4
Expected	5.0	12.0	14.4	11.5	6.9	5.2

So the value of $\Sigma \frac{(O-E)^2}{E}$ is $(0.2 + 0.0833 + 0.1361 \ldots) = 1.241$.

This is the number of frequency cells less 1 for the totals to agree and 1 more because the mean of the Poisson was derived from the data.

The number of degrees of freedom is $6 - 1 - 1$.

The frequency cells are reduced to just 6 and the mean has had to be calculated from the data.

The critical value is 9.488. The test statistic is much less than this, hence the model is appropriate.

Q3 The null hypothesis is that the cover and content are independent, the alternative hypothesis is that they are associated. This is a 2×3 table and so there are two degrees of freedom, the critical value is 5.991.

State your hypotheses even though they are not specifically asked for.

State the critical value.

The expected values are:

	Reference	**Light**	**Serious**
Paperback	30	41.1	38.9
Hardback	24	32.9	31.1

The test statistic is 10.80 +…

If individual cell statistics exceed the critical value do not carry on calculating – you have done enough.

Since this is already greater than 5.991 it is possible to conclude that this is a significant result. This means that the null hypothesis is rejected. Thus a book's cover and its content are associated.

Q4 The null hypothesis of independence leads to the expected frequencies given in the following table.

Age	**Theft**	**Licence**
Under 20	19.2	28.8
20 and over	20.8	31.2

Yates' correction involves finding the values of
$\frac{(|30 - 19.2| - 0.5)^2}{19.2}$
$+ \ldots = (0.5672 + \ldots + 0.3490) = 1.818$.
The 5% critical value is 3.841, hence the null hypothesis is accepted, therefore age and crime are independent.

17 Hypothesis testing

Q1 $H_0: p = 0.35$, $H_1: p < 0.35$

In B(20, 0.35): $P(X \leqslant 4) = 0.1182$ which is not significant, hence there is insufficient evidence that these are more reliable.

Here, the situation described may be modelled using a binomial: B(20, 0.35), so the hypotheses are appropriate, since greater reliability would lead to fewer computers developing faults.

Q2 Use the Poisson distribution, Poi(8).

$H_0: \mu = 8$, $H_1: \mu > 8$ so the upper tail is of interest.

$P(X \geqslant 14) = 1 - P(X < 14)$
$= 0.0342$

On this basis there is evidence of a higher radiation level.

This is a classic random events in a continuous medium context, hence the Poisson distribution, Poi(8), may be used as the underlying probability model.

Q3

a) $H_0: \mu = 8.5$, $H_1: \mu < 8.5$

The z-statistic is $\dfrac{8.02 - 8.5}{\frac{0.51}{\sqrt{10}}}$

$= -2.976$ which is significant compared with -1.96.

Thus there is evidence to suggest that the mean is less than 8.5 kN.

b) The mean breaking strength is 8.5 kN but we conclude that it is less than 8.5.

The likelihood of this is 2.5%.

Q4 $r = 0.8680$ compared with the 2.5% critical value 0.7067, hence there is evidence that the weights are positively linearly related.

Note the indicative word 'linear' which tells you to use Pearson's coefficient.

Q5 Replace the literal scale with ranks where 1 corresponds to A etc. Then Spearman's coefficient for these data, r_s, is 0.4286 compared with the 5% critical value 0.7143, hence there is insufficient evidence to support the claim.

There are no tied ranks, hence either approach gives the same result.

Q6

a) $H_0: \pi = 0.75$, $H_1: \pi < 0.75$.

b) $z > 2.054$

c) $p \sim N(0.75, \dfrac{0.75 \times 0.25}{40})$,

and the z-statistic here is

$$\frac{\frac{36}{40} - 0.75}{\sqrt{\frac{0.75 \times 0.25}{40}}} = 2.191$$

hence this result is significant.

The drug company is justified in its claim based on these data.

18 Approximating distributions

Q1 Using Poi(6): $P(7) = 0.138$

Here, the situation described may be modelled using a binomial: B(120, 0.05). The probability is reasonably small, and the mean is not particularly large so a Poisson approximation seems reasonable.

Q2 $P(X = 26)$ becomes $P(X \leqslant 26.5) - P(X \leqslant 25.5)$

$= \Phi\left(\dfrac{26.5 - 25}{5}\right) - \Phi\left(\dfrac{25.5 - 25}{5}\right)$

$= 0.078$

The Poisson distribution, Poi(25), applies but the mean is large. Hence N(25, 25) may be used to approximate the probability.

Q3 $P(X > 400) \approx P\left(z > \dfrac{400.5 - 380}{\sqrt{235.6}}\right) = 0.091$

This is a binomial context: B(1000, 0.38) which can be approximated using the normal: N(380, 235.6).

Q4 $N(20 \times 10.52, 20 \times 4.2)$ is appropriate, hence $P(X \geqslant 200)$

$= 1 - \Phi\left(\dfrac{200 - 210.4}{\sqrt{84}}\right)$

$= 0.872$.

19 Continuous random variables

(a) $\int kx(4-x)^2 dx = 1$

This gives $k = \frac{3}{64}$.

The value of the integral under the density function over the interval must be unity (1).

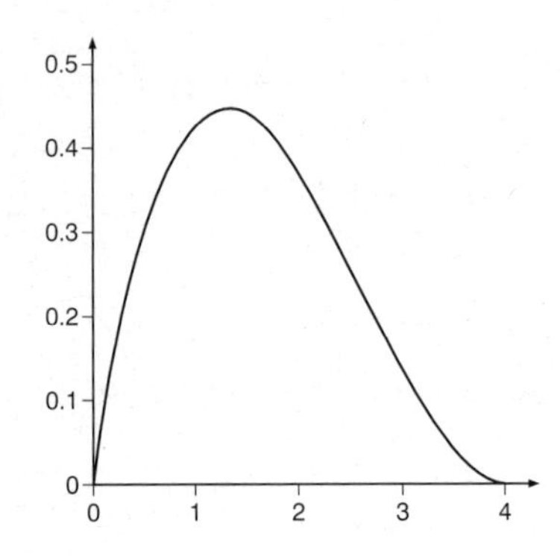

(b) The mean is from

$\int kx(4-x)^2 dx = 1.6$

Expected value is the technical terminology for the mean.

(c) Mode = $\frac{4}{3}$

The sketch indicates that the modal value occurs at the calculus maximum.

(a) $\int kx^{-2} dx = 1$, hence $1 = -k[1 - 10]$, giving the result $k = \frac{1}{9}$.

(b)
$$F(x) = 0 \quad \text{for } x < 0.1$$
$$= \frac{10 - x^{-1}}{9} \quad \text{for } 0.1 \leqslant x \leqslant 1.0$$
$$= 1 \quad \text{for } x > 1.0$$

(c) $P(x > 0.5) = 1 - F(0.5) = \frac{1}{9}$

(d) $F(u) = 0.5$, therefore $x = \frac{2}{11}$

$F(q_1) = 0.25$, therefore

$q_1 = \frac{4}{31}$ and similarly $q_3 = \frac{4}{13}$

(a) For a distribution function, F(minimum) = 0 and F(maximum) = 1.

Thus with $a = 2$ and $b = 8$, $F(2) = 0$ and $F(4) = 1$.

(b) $f(x) = F'(x) = \frac{1}{8}(10 - 2x)$

(c) $F(m) = 0.5$, therefore $\frac{1}{8}(m-2)(8-m) = 0.5$.

This leads to the median equalling $5 - \sqrt{5}$.

(d) Mean = $\int x f(x) dx = \frac{17}{6}$

(a) Mean = $\int \frac{24}{t^2} dt = 4$

Therefore the variance

$= \int \frac{24}{t} dt - 4^2 = 24 \ln 2 - 16$

$= 0.635\,53$.

Hence the standard deviation is 0.7972.

(b) $F(t) = 0$ for $t < 3$ and

$\int \frac{24}{t^3} dt = \frac{4}{3} - \frac{12}{t^2}$

for $3 \leqslant t \leqslant 6$ and 1 for $t > 6$.

(c) $P(2 < t < 5) = F(5) - F(2)$

$= \left(\frac{4}{3} - \frac{12}{25}\right) - 0 = \frac{64}{75}$